The Otter and the Fairy Shrimp

James H. Thorp

The Otter and the Fairy Shrimp

 Springer

James H. Thorp
Department of Ecology and Evolutionary Biology
Kansas Biological Survey and Center for Ecological Research
University of Kansas
Lawrence, KS, USA

ISBN 978-3-031-64028-5   ISBN 978-3-031-64029-2   (eBook)
https://doi.org/10.1007/978-3-031-64029-2

This Springer imprint is published by the registered company Springer Nature Switzerland AG
The registered company address is: Gewerbestrasse 11, 6330 Cham, Switzerland

If disposing of this product, please recycle the paper.

About the Author

The Author: James Thorp often claims that he is considerably "older than dirt but still kicking"! In his wandering life, he has received a master's degree for freshwater studies of green hydra from North Carolina State University (NCSU) and a Ph.D. from NCSU with a dissertation focused on marine biology and estuarine grass shrimp. Most of his doctoral courses were taken at the Duke University Marine Laboratory, but shortly afterwards he migrated upstream to freshwater studies. After serving as a Dean of Science, Chair of Biology, and Field Station Director at various universities, he returned in 2001 to his undergraduate *alma mater* (University of Kansas) as a tenured Professor and Senior Scientist. He has conducted freshwater research in fields ranging from physiological, behavioral, and community ecology, all the way up (spatially) to macrosystem ecology in multiple US states and even continents. Although he has published studies of fish and amphibians, his focal creatures have often been invertebrates. In fact, he has more than a dozen books authored or edited on freshwater invertebrates from around the world. James and his wife Margaret live in an 1860s stone house in Lawrence, Kansas.

Acknowledgment

This book would have suffered significantly without the help of my colleague Christopher F. Frazier, an aquatic ecologist and photographer who was born in southwestern Michigan. He provided some original photographs and tracked down other free ones for our use from the web. Christopher was first enamored by freshwater animals after being introduced to Great Lakes zooplankton communities while earning his B.S. at Central Michigan University. His work then led him to earn his M.S. at Western Michigan University studying aquatic insect communities in Lake Michigan's coastal wetlands. Since finishing his graduate education, he has been involved in research projects spanning Michigan wetlands, Texas coastal streams, and ephemeral wetlands across the North American Great Plains region through the University of Kansas. All those studies have been primarily focused on aquatic invertebrates. While sharing the mysteries of the natural world via research has been great, he found in 2019 that high-quality images typically evoke much more emotion from the public than graphs typically do and has since been an avid photographer to spread his love of wildlife. Naturally, his favorite subjects are those that live in the water: aquatic invertebrates, waterfowl, wading birds, and aquatic and semi-aquatic mammals, but he also rarely declines a chance to take a few photos of charismatic terrestrial animals!

About This Book

Most of us encounter rivers, lakes, or wetlands on a regular basis even if from a distance, but very few of us know more than a few facts about the organisms that inhabit them and what ecological factors control their lives. One purpose of this book, therefore, is to begin to rectify that information deficit for a broad range of the general public from teenagers to retirees. Another goal is to inform the public about aquatic systems in the hope that this will eventually produce an electorate better able to judge issues on the ballot or in community meetings that might impact these ecosystems. Although my normal audience over the decades as a professor and aquatic ecologist has been undergraduates, graduate students, and fellow scientists, I have tried to write this book using terms more familiar to the general public and in a manner that reveals many of the exciting aspects of freshwater ecology. Most chapters include information from around the world, though the majority is naturally drawn from ecosystems and scientists in the northern hemisphere where I have more field experience. Because the audience for this book is likely to be international, I have employed both metric units (e.g., kilograms and meters) and imperial units (e.g., pounds and feet).

This book is organized into ten chapters, with six on organisms and four on the aquatic systems where they live, such as lakes, rivers, caverns, and bogs. I have tried—with only partial success—to avoid repeating the same information in multiple chapters, but this is a very challenging task when it comes to aquatic organisms that live in multiple habitats and have diverse attributes worthy of discussion from multiple perspectives. Chapter 1 focuses on aquatic mammals, primarily otters. Fairy shrimp and a large diversity of freshwater invertebrates are described in Chap. 2. Aspects of the lives and times of fish are discussed in Chap. 3, though this topic is also covered in other ways in Chaps. 4 and 6. Behemoths of the freshwater world—from hippos to giant snakes—are the subject of Chap. 4. The theme of Chap. 5 is aquatic architects with coverage of some organisms that you would expect—like beavers—but also with some unexpected topics like the nature of "houses" made by caddisfly insects. Most of Chap. 6 is focused on freshwater species that are usually considered denizens of the ocean, including sharks, rays, dolphins, seals, and manatees. From that major six-part section of the book, the focus then shifts to a four-part analysis of the physicochemical and ecological attributes of aquatic habitats including lakes (Chap. 7), rivers (Chap. 8), underground aquatic systems (Chap. 9), and wetlands (Chap. 10). Having a better understanding of the nature of these ecosystems should provide readers with critical information on the ecological advantages and disadvantages for the existing and potential residents of those systems.

Contents

Chapter 1
Otterly Wonderful

Or should that be "utterly" wonderful? Well, the answer will depend somewhat on your perspective because each of us examines life through a different lens. Many readers will focus on the unique attributes of this very intelligent and playful mammal, but others may be annoyed by the havoc river otters can sometimes cause whether interacting with city dwellers—as in Singapore—or competing with anglers on a fishing trip, but more on those aspects later, and let's start with some basic scientific observations.

Will the Real Otter Please Step Forth?

Sorry, "no can do" … because our globe supports 13 species of otters on 6 continents, all of which are members of the scientific family Mustelidae. That large taxonomic family of mammals in the order Carnivora also contains many other less publicized species, such as badgers, ferrets, martens, minks, weasels, and the large and dangerous wolverines. Because of its close contact with residents in coastal cities of the northern and eastern North Pacific Ocean and its star coverage on many television documentaries, the marine sea otter (scientific name = *Enhydra lutris*) has garnered the most attention from the general public. In contrast, the more petite, semiaquatic river otter (*Lontra canadensis*) of North America (Fig. 1.1) is more widespread geographically but tips the scale at only 11–31 lb (0.5–14 kg). While the North American river otter is still impressive in size when viewed in the wild, its largest relative is the giant river otter (*Pteronura brasiliensis*) of South America, which weighs 49–57 lb (22–26 kg). Because of this book's nature, of course, the river otter is an important focus of this chapter. Other otter species are well-known to the public in some countries, including urban populations of two species of otters in the city nation of Singapore: the smooth-coated otter and the Asian small-clawed otter, as discussed later in this chapter. Many urban otters are undergoing important

J. H. Thorp, *The Otter and the Fairy Shrimp*,
https://doi.org/10.1007/978-3-031-64029-2_1

Fig. 1.1 Close-up view of the river otter *Lontra canadensis* shown eating a fish at Baker Wetlands near Lawrence, Kansas, USA. (Photograph courtesy of Christopher F. Frazier)

behavioral modifications while adjusting to humans and the opportunities and constraints associated with living in close proximity to us and our pets. Before investigating urban populations with their human-otter interactions, let's learn more about the North American river otters in the wild.

Where or Where Can My Otter Be?

To avoid confusing some of my readers at the start of this chapter, I should explain two facts. First, the term "river" otter is a misnomer because they are not confined to rivers! Indeed, these mammals inhabit freshwater streams, ponds, and wetlands from the mountains to the sea. Moreover, in rare occasions, they can be observed in the upper regions of estuarine habitats along marine coasts. However, don't waste much of your time searching for them there and certainly not in the kelp beds of the US West Coast where their marine otter relatives frolic and dominate the scene. Second, the original range of river otters included much of North America where natural, relatively permanent aquatic systems existed. In modern times, however,

Fig. 1.2 Side view of *Lontra canadensis*. (Photograph courtesy of Flickr contributor Thomas Koerner at the US Fish and Wildlife Service)

urbanization, pollution, and hunting have constricted their range mostly to rural Canada and the USA. The latter includes coastal areas, a few midwestern states, and mountainous regions where human disturbance is minimal and water is plentiful. They have almost disappeared from much of the Great Plains states except where otter families (or "romps") have been reintroduced. However, they are occasionally spotted in the Kansas River and in the wetlands on the southern edge of Lawrence, KS (my hometown) (Fig. 1.2). In some rare cases, these reintroductions to central and western states have been so successful that trapping seasons have been re-implemented—much to the dismay of most nature lovers who are shocked that we continue killing any animal species for its pelt.

Otters are rarely spotted by humans even though they occupy many aquatic habitats because they are mostly nocturnal or at least crepuscular. They spend much of their time—especially during daylight hours—within their terrestrial homes, which are typically located where sufficient cover exists from trees, other abundant vegetation, and/or large rocks to provide some protection when they are foraging. Less desirable to the average otter is being forced to build a lodge under a fallen tree or other debris where exposure to predators is greater. Their burrows are lined with grass, leaves, moss, and other forms of dead vegetation that are suitable for padding and warmth. Their homes are always located close to a relatively permanent water body and may consist of a new dwelling or a burrow previously occupied by another mammal species, such as a beaver, fox, muskrat, or woodchuck. The otter is wise enough to have, if at all possible, at least two openings to its den (or "holt"), one of which preferably opens underwater. Daylight activity tends to increase somewhat in winter months when temperatures are at least minimally adequate and potential predators are less active in general.

Menu Preferences and Food Acquisition

Otters have a rather eclectic diet focused entirely on live animals—with menu selections based on their personal inclination for what looks tasty that day and the prey's relative availability and difficulty of capture. Fish and crayfish often top the list of preferred consumables (Fig. 1.3), but the relative proportion of their diet varies with the season of the year and the local availability of items that can realistically be captured. The cuisine is also known to periodically include clams and snails, aquatic insects, frogs and salamanders, small rodents, turtles, and wading or molting birds that are temporarily incapable of aerial escape. Even a small dog might be attacked and eaten if it strays into the water and is drowned by the otter! Small prey can be eaten while the otter floats in the water, but larger prey are usually first taken to shore. It is not unusual to see two or more otters cooperating to capture larger prey.

To acquire aquatic food, otters must obviously be adept at swimming and locating prey (Fig. 1.4). Fortunately, their relatively long tails (nearly 40% of their body length) and powerful hind feet are ideal attributes for this activity. Indeed, they are remarkably fast swimmers at nearly 7 mph/ll kph and moderately deep divers (66 ft/20 m), with the capacity to stay underwater for up to 4 min. The underwater time must obviously include time to locate and capture the victim and then return to the surface. In comparison with the otter's underwater velocity, the fastest human

Fig. 1.3 Young river otter eating a fish. (Photograph courtesy of Flickr contributor "marneejill")

Fig. 1.4 Swimming otter from Baker Wetlands. (Photograph courtesy of Christopher F. Frazier)

swimmer has reached 6 mph/1.6 kmh in Olympic competition. Acquiring and retaining prey are aided by the notable tactile sensitivity of their facial whiskers and the agility of their paws to grab and hold the prey. The otter's well-refined senses of hearing and smell greatly enhance prey detection and locating potential predators in both day and night. Their visual attributes have evolved in ways that have allowed them to acquire aquatic prey, including possessing relatively transparent, nictitating membranes to protect their eyes underwater. When submerged, otters close their nostrils and ears but keep their eyes open. Their evolved visual aptitude for underwater hunting comes at a cost, however, as that visual acuity is accordingly reduced when the otter peers above the water or is on land. Their noted near-sightedness when above water may be responsible for periodic observations that otters will approach fairly close to a boat occupied by humans before retreating to safe environs upon realizing their potentially dangerous error.

Rapid and moderately long swimming events are possible because otters have evolved the requisite body structures and senses, including the eyes that are perched high on their head (Fig. 1.5). Their slim head and tapered, streamlined body lower underwater drag and reduce energy expenditures, thereby allowing increased swimming speed. Their swimming power is also enhanced by short legs, webbed feet, and a long powerful tail. Also contributing to their aquatic abilities is their dense undercoating of hairs which traps air for insulation and which is itself overlain by a layer of long water-resistant hairs. This body structure makes them ideal freshwater mammals (though slower than freshwater dolphins), but it is less desirable for terrestrial locomotion. Nonetheless, a river otter can run on land at speeds up to

Fig. 1.5 Close-up view of
the head of an otter.
(Photograph courtesy of
Flickr contributor from the
Maritime Aquarium at
Norwalk)

15 mph/24 kph—though only for short distances. By comparison, the fastest human
has been clocked at almost 28 mph/45 kph—also for very brief periods.

Social Structure and Interactions

When not hunting or playing, an individual or pair of otters and their pups, if present, typically occupy a burrow, which is also known as nests, dens, holts, or couches.
They frequent their dens during daylight hours when they would otherwise be more
exposed to predators. Ideally this burrow has a primary underwater entrance to
maximize protection for the adults and young when entering or leaving their lodgings. The burrow resides in a home range of highly variable size depending on the
otter's gender, nature of the environment, presence of other otters, a female's reproductive stage, and frequency of human interactions. Home ranges typically range
from 3 to 15 sq. mi (7.8–39 sq. km) and are as large as 30 sq. mi/78 sq. km, with
smaller ranges being more common. Home ranges of males tend to be larger than
those of females.

Otters become reproductively active at an age of 2–3 years. They most often
mate in the water in late winter or early spring, with a gestation period typically
lasting 9.5–12.5 months. Pregnant females commonly stay alone in their burrows
until the young are born. A litter most often contains one to three "pups or kittens"
(Fig. 1.6), but rarely up to six have been noted. Baby otters are born blind and are
highly dependent on their mothers for the first 2 months before they can leave the
burrow temporarily to roam the adjacent pond or stream. They may depart for an
independent existence after 6 months or stay with their mother for up to a year. A
typical adult will survive up to 9 years in the wild, but individuals have lived twice
as long or even slightly more in zoos.

Otters are a social species even though they may nest individually. Indeed they
are usually considered the most social animals within their taxonomic family, which
as stated earlier includes badgers, ferrets, martens, minks, weasels, and wolverines.

Fig. 1.6 Baby otters. (Photograph courtesy of Flickr contributor David Brossard)

Family groups of related individuals are common, and unrelated males may form a temporary "social pack" (= bevy, lodge, or romp) with more than a dozen members and collaborate in hunting, grooming, and even den sharing. Separate groups of otters tend to avoid each other, especially during cooler seasons. Males tend to maintain larger home ranges than those of females.

Inter- and intraspecies interactions are primarily based on a repertoire of vocal communication that includes whistles and a variety of growls and loud screams—the latter can sometimes be heard from more than a mile away. They also secondarily employ tactile interactions and messages expressed silently by body posture. Territories are typically marked by potent, musky chemicals from scent glands that are primarily centered near the base of their tails, though their urine and feces also facilitate this communication. Young otters develop these glands as they mature, and the released scent differs between males and females.

Urban Otters

My recipe for wild mammals to be at least temporary success in metropolitan havens is to be "cute and photogenic, environmentally adaptable, family oriented as a species, and able to effectively exploit the habitat for food and shelter"! It also does not hurt if you can indirectly manipulate the media so you are a star of Internet sites like Facebook and fan clubs, as has happened for otters in Singapore! Oh yes, and don't cause too many urban problems that would enable your critics to outnumber your supporters.

Fig. 1.7 Asian small-clawed otter (*Aonyx cinereus*). (Photograph courtesy of Flickr contributor Mathias Appel)

Fig. 1.8 Smooth-coated otter (*Lutrogale perspicillata*). (Photograph courtesy of Flickr contributor Manan Singh Mahadev)

All these characteristics currently apply to the two species of otters now living in the city nation of Singapore (Fig. 1.7) on the Malay peninsula: the Asian small-clawed otter (*Aonyx cinereus*, formerly called *Amblonyx cinereus*) and the smooth-coated otter (*Lutrogale perspicillata*) (Fig. 1.8). They are the two most common otter taxa among the four species reported from mainland Malaysia, with the other two being the "common or Eurasian otter" (*Lutra lutra*) and the "hairy-nosed otter" (*Lutra sumatrana*). All four otter species are listed in the "threatened species category" of the IUCN Red List of Threatened Animals (UUCN = International Union for Conservation of Nature). The Eurasian otter is designated as "vulnerable" on that list, while the status of each of the other three species is classified as "insufficiently known" because of inadequate population and distribution data for these otters in the wild.

Otters—including the two Singapore species—compete among multiple otter species while also facing native Asian predators that include crocodiles, huge snakes, monitor lizards, and jungle cat species. But, when two otter families or "romps" colonized Singapore's urban environment from surrounding natural habitat late in the last century, their main "predators" changed to automobiles (less than ten deaths per year) along with traditional predators in the lakes of Singapore, such as crocodiles. Their population size had increased from 90 to 150 in less than a decade (as of 2022), with most being the smooth-coated otter. However, the number of families has stayed relatively steady around ten, but with an increase in number of members per romp—a trend that is uncommon in completely wild populations. For example, the notable "Bishan Otter Family" had ten members in 2016. The territory requirement of families may ultimately limit the population size with the urban environment, though they could continue to grow in more rural areas of this nation.

Urban coexistence has required some adjustments in lifestyle, just as it does in different ways for humans moving from the country into cities. Otter pups now necessarily stay within the romp longer than they do in the wild. They still prefer to eat fish—but now that includes species in local reservoirs and rivers as well as highly prized koi in tended private ponds of Singapore residents (Fig. 1.9). As you can well imagine, this latter tendency has led to an understandable uproar from

Fig. 1.9 Koi pond in Singapore invaded by hungry otters. (Photograph courtesy of Flickr contributor Tanay Kibe)

those humans who have long tended these expensive fish that can cost from $10 for small juveniles up to $50,000 each for champion quality koi! Aside from challenges of competition among otter families, dodging cars, and finding adequate food, another hurdle for these urban otters is acquiring suitable dens. Those typically available in the wild have largely disappeared in Singapore except along the shoreline of local lakes and rivers, and thus some urban otter families have turned to denning below ground in human canals, under bridges, and in grassy open spaces of the city. Drainpipes also provide urban highways for the otters, while city fountains provide a ready source of drinking water. Interactions with humans have largely been positive as long as people do not try to grab them or disturb their pups. When the urban otters are sufficiently disturbed, they will yelp, whimper, or scream to rally their family members and will bite when sufficiently provoked. Remember if you observe otters in either a natural or urban environment that they are still wild animals with an instinct to protect their young.

To learn more about the urban otters of Singapore, seek fascinating information from the web and even from dedicated Facebook pages developed by human residents of Singapore!

Chapter 2
Fairies and Other Small Hidden Creatures

Whole books are published on invertebrates in general, on insects only, and on aquatic invertebrates in general. In this chapter, however, I will focus on only a few representative groups that I hope may be of particular interest to many readers. I hope this coverage will spur your desire to collect some specimens and learn first-hand more about the diverse world of freshwater invertebrates. Most of these can be observed and/or collected locally, unless otherwise indicated, and photographs of these are easily found on Wikipedia and other Internet sites. These organisms are critical for the transfer of energy up the animal food web from both aquatic primary producers (cyanobacteria, true algae, mosses, and aquatic weeds) and terrestrial vegetation that has flowed into an aquatic system. Without these vital invertebrate consumers, survival would not be possible for the larger and more familiar aquatic animals, such as fish, turtles, and mammals. In addition to reading about these organisms in the present chapter, please also consult other chapters in this book for more information on animals with and without backbones. For example, Chap. 5's discussion of invertebrate architects (crayfish burrowers and caddisfly case and retreat makers) may be especially relevant and could spark your exploratory nature and lead to new adventures.

Large exotic animals often rate special features on television, but we tend to overlook and ignore the more common but minute creatures that surround us. Many invertebrates are hard to find initially but easy to see with the naked eye once you locate them, while others are easy to find but hard to see without a microscope or a strong magnifying glass. The latter include both single-celled protozoa (= Protista) and many small multicellular animals that live in all non-thermal freshwaters:

> As a side note for all parents and other relatives, please buy your child a simple microscope or strong hand lens while they are still in late grade school, as this may open new vistas for them and might be the spark that eventually produces a future scientist … as it did for me as a child in grade school many decades ago!

J. H. Thorp, *The Otter and the Fairy Shrimp*,
https://doi.org/10.1007/978-3-031-64029-2_2

The current chapter covers a large diversity of freshwater invertebrates and ends with the exploration of large branchiopod crustaceans, including clam, tadpole, fairy shrimps, etc., also known by many children from the "sea monkey" eggs in most pet stores that also sell fish, aquaria, and other aquatic supplies.

Microorganisms

This section focuses on a wide diversity of microscopic organisms from single-celled protozoa to really tiny multicellular creatures, some of which are related to much larger species discussed later in this chapter. Scientists classify these organisms into a wide diversity of phyla, but I am grouping them here for the simple reason that most require a microscope or strong magnifying lens to view them, and thus many readers will not otherwise seek to find them.

Single-Celled Organisms: Protozoa and Other Protista

Scientists have often disagreed over the decades and centuries on how to classify single-celled organisms, with their very simple body organization. We know they are not fungi nor multicellular animals, and they cannot be classified as plants even though some may depend heavily on internal photosynthesis. Protists were formerly grouped as either protozoa or algae, with only the latter capable of photosynthesis. We now know that the distinction is both narrower in some cases and wider than formerly recognized in others. And as a result, scientists classify them taxonomically into multiple phyla and sometimes even into different kingdoms, with some groups capable of photosynthesis and others being more "animallike." They occur in numerous wet places, including damp soils, freshwaters, and the oceans. Many play a vital role in transferring energy to higher forms of aquatic life. While the vast majority are neutral or beneficial to humans, some are associated with human diseases, such as malaria which, as many of you know, is transmitted by some species of mosquitoes. Photographs of many types of protists are easily viewed on the web.

To best observe aquatic protists, collect separate samples of water from near the surface of ponds, wetlands, and backwaters of streams or from the bottom of the water body using a pipette or turkey baster while avoiding stirring up silt. You can also extract a small amount of aquatic vegetation and swish the plant material in a jar as a potential source of these critters. If the collected water is murky, then let it settle in the jar before using a medicine dropper to extract a sample to place on a microscope slide to look for these protists.

Hairybacks: Phylum Gastrotricha

The multicellular gastrotrichs (Fig. 2.1) might be confused with giant hairy proto-zoans (hence the names hairybellies or hairybacks), as these "behemoths" are in length only about 0.1 in/0.25 cm and are structurally rather simple. Most crawl on or barely within the sediment of the water body or along the underwater surface of plants. Though very tiny, they have a far more efficient, one-directional gut (mouth to anus), which makes them significantly more advanced than protists. Gastrotrichs are mostly fork-tailed, bottom-feeding animals that eat both dead organic matter and living bacteria, algae, and small protists.

Seed Shrimps: Phylum Ostracoda

Ostracods are benthic (freshwater and some marine) and planktonic (marine mostly) crustaceans that can only be seen well with a microscope, but they can make their presence known in the ocean with the blue-green lights that individuals can pro-duce—a technique generally rare in freshwater species. If you collect a small amount (0.13 in/0.32 cm) or less depth of a sandy-silty sediment from a clean, oxy-genated lake or backwaters of a river that contains sufficient organic material, you may find these small (~0.04 in/0.1 cm), multicellular animals with their bivalved, chitinous shells (Fig. 2.2). These "seed shrimps" or "mussel shrimps" can be easily confused with branchiopod clam shrimps (discussed later in this chapter), as the bodies of both are encased in a bivalved covering. Their diets consist of minute organisms like algae, diatoms, bacteria, molds, and organic detritus present on the lake/stream bottom or on the surface of vegetation.

Fig. 2.1 Microscopic gastrotrich (common name: hairybellies or hairybacks). (Photograph courtesy of Flickr contributor Philippe Garcelon)

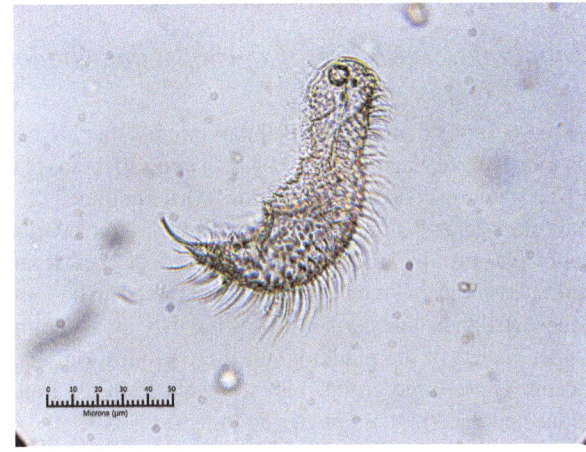

Fig. 2.2 Microscopic seed
shrimp (Ostracoda).
(Photograph courtesy of
Flickr contributor Philippe
Garcelon)

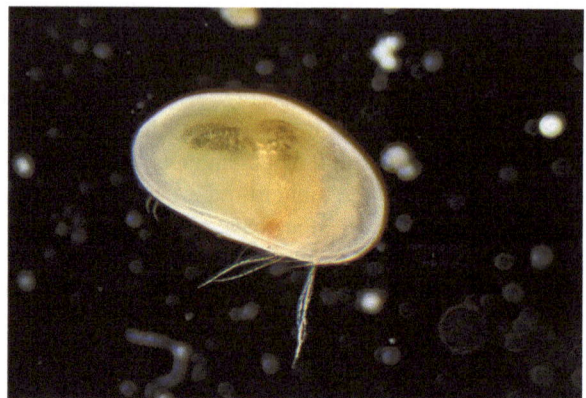

Fig. 2.3 Microscopic
rotifer (*Brachionus
quadridentatus*), or "wheel
animal." (Photograph
courtesy of iNaturalist
contributor Ken Kneidel)

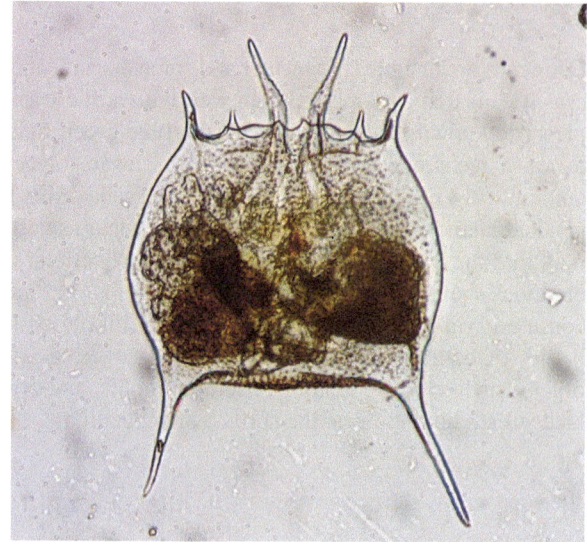

Zooplankton: Rotifers, Copepods, and Cladocerans

Three groups of small multicellular animals that are extremely easy to collect with
an inexpensive, small mesh (0.04 inch pores; 0.1 cm) plankton net and then observed
under a microscope are rotifers (or "wheel animals"; Fig. 2.3), copepods (Fig. 2.4),
and cladocerans (Fig. 2.5). Rotifers are in their own phylum (Rotifera), while the
latter two groups are in separate classes of crustaceans in the largest phylum of
animals on Earth—the Arthropoda. All three can be found near the bottom of ponds,
lakes, wetlands, and parts of rivers, but they most commonly occur suspended in the
upper waters of the plankton. Of these groups, only copepods are abundant in the
ocean. These three groups play a huge role in the transfer of organic energy from
algae and dead organic matter up to fish.

Fig. 2.4 Ventral view of a marine calanoid copepod; the locomotory anterior antennae are actually two to three times longer than shown here, and adults are ~ 0.7 in/17 mm long. (Photograph courtesy of Flickr contributor Russ Hopcroft from the Gulf of Alaska Seamounts 2019 Expedition)

Fig. 2.5 Lateral view of a water flea (Cladocera, *Ceriodaphnia dubia*). (Photograph courtesy of iNaturalist contributor Ken Kneidel)

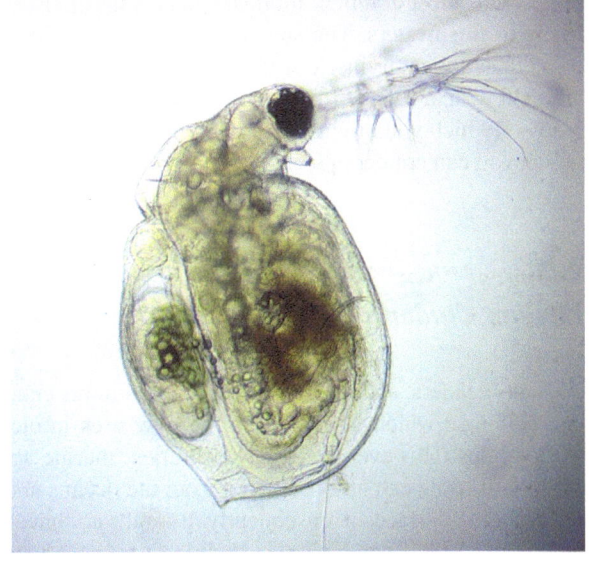

Rotifers—or "wheel animals"—are named for their crown of cilia (called the corona) which gives the false impression in living animals of actually whirling around. This ciliated crown both propels these minute zooplankton through the water and sweeps food into the gut of bottom-dwelling (benthic) and suspended (planktonic) rotifers. The largest rotifers are only about 0.08 in/0.2 cm long. Many taxa reproduce sexually; but oddly enough, the bdelloid rotifers reproduce entirely by asexual means.

Crustacean zooplankton are larger than rotifers but less abundant, and both copepods and cladocerans swim using jerking motions of their large anterior appendages. Copepods are ubiquitous species that are especially dominant in the plankton and are only 0.02–0.08 in (0.05–0.2 cm) long. While some predaceous species exist (especially cyclopoid copepods), most feed on algae in both freshwater and marine habitats. Copepods can be found in both the water column and on the bottom in almost all freshwater lakes and streams. Cladocera are sometimes called "water fleas," though the name is entirely misleading, given that true fleas are terrestrial insects! Most are microscopic (0.02–0.7 in; 0.02–1.8 cm), but *Daphnia magna* is large enough to be seen clearly when swimming in a jar of lake water. Again, most are algivores. Fish are major predators of both cladocera and copepods.

Macroorganisms: Simple and Complex

In addition to the microscopic, multicellular animals described above, most freshwater invertebrates are much larger and many are more complex. They range from structurally and physically simple organisms—like sponges and hydras—to the largest and most complex mollusks (snails and mussels), crustaceans, and insects living in freshwaters. The smallest of those described below can be seen with no more than a magnifying lens if necessary, but most require no supplemental visual aid. A few taxa selected from the vast diversity of this group are described below, many of which you might see in a science lab in high school or college and most of which you can collect in local streams and ponds.

Sponges: "Not" What You Buy in the Grocery Store— Phylum Cnidaria

To many readers, sponges are manufactured items constructed of polyester or vegetal cellulose which almost everyone now uses in cleaning items from dishes to automobiles. However, for many centuries, marine sponges were regularly—and still are to some extent—harvested from the oceans and Mediterranean Sea for use in scrubbing surfaces. Consequently, it might not have dawned on readers who are in the post-WWII generations—or at least to those who are not scuba divers—that sponges are actually very simple, colonial animals in the phylum Porifera. Indeed, they are one of the least complex, multicellular organisms in the animal kingdom, despite their frequently large size. They have a cellular level of organization characterized by the lack of true tissues and organs and instead depend on omnipotent collar cells (= choanocytes) to move water and food from outer pores into interior canals and out one or multiple chamber openings. While all sponges lack a true skeleton, their bodies are usually supported by tiny, spiky, silica spicules and a noncellular matrix called "spongin." Marine sponges must be processed to remove these spicules before they reach the commercial market, but freshwater sponges are not suitable in size or shape for human use. Despite their structural simplicity, marine sponge colonies can be quite massive; for example, oceanic barrel sponges can exceed a human's height!

Fig. 2.6 Freshwater sponge (family Spongillidae) on a small rock. (Photograph courtesy of iNaturalist contributor Edward Hicks)

Even for those of you who have seen marine sponges, it might not have occurred to you that freshwater sponges (Fig. 2.6) can be found in clean streams, though more than 97% of all sponge species are marine. The very simple and small freshwater sponges occur in colonies on the top of or partially under rocks where they live as low encrusting to relatively massive spreading or arborescent colonies. However, they are not and could not be commercially exploited because of their small size and physical structure. The freshwater species come in various colors, but primarily black, brown, cream, white, or green. The green forms have symbiotic algae enmeshed within the sponge's noncellular matrix and thus require exposure to rays of the sun so that the algae can grow and provide energy to the sponge. Be cautious when seeking them, however, because this green color makes it easy to confuse these animals with algae. All freshwater sponges are also filter feeders on bacterioplankton and dead organic matter. They are eaten by a few aquatic insects, including by specialized spongilliaflies, as well as by crayfish and even ducks.

Hydra and Freshwater Jellyfish: Phylum Cnidaria

Freshwater members of the almost exclusively marine phylum Cnidaria are characterized by substantial increases in body complexity over protists and sponges
because they are not only visible to the naked eye but also possess two cellular layers (outer and inner ectoderm)—a significant improvement over a simpler cellular
level of construction. While they have only one opening to their digestive cavity,
they are still members of a very successful phylum of animals that includes marine
corals and jellyfish species. However, fewer than ten freshwater species of hydra are
found worldwide, but that is still more than their freshwater, jellyfish-looking relative, *Craspedacusta sowerbii* (see description below). Despite their diversity and
ecological importance in marine habitats, however, cnidarians are often considered
a "dead-end" phylum because they seem not to have given rise over evolutionary
time to more complex animal phyla.

Hydra are rather simple, multicellular animals that possess 1–12 tentacles
(Fig. 2.7), with low numbers being more common. Each tentacle contains stinging
cells known as cnidocytes (or nematocytes) equipped with special triggers that
release a neurotoxin. The paralyzed invertebrate prey is drawn into the central,

Fig. 2.7 Light-colored
green hydra. (Photograph
courtesy of iNaturalist
contributor Eamon Corbett)

gastrovascular cavity where it is digested with extracellular enzymes over many hours. No hydra or any other cnidarian possess a neural structure that we would call a brain. Related to their simple construction is their ability to regenerate body parts relatively easily and frequently. While locomotion can obviously be a challenge for an animal lacking legs, hydras can slowly move by amoeboid action at the base of the body column (more-or-less like scooting across the substrate), and they can even somersault using the entire body, or they can release from the solid surface and drift with the water current!

Most of a hydra's day is spent attached to an underwater plant, wood, or rock where they wait to attack passing zooplankton prey by using their extrusible sting-ing cells—all of which are too small to hurt people. While waiting for passing prey, hydra continue reproducing with asexual buds that are easily visible on the hydra's outer surface and may contain tentacles of their own. If environmental conditions are poor, however, they can also reproduce sexually with one individual releasing a male gamete that may eventually find its way to the ovary of a nearby female. In one sense, they are almost ageless, because they can continue budding and replacing body parts in perpetuity unless they succumb to outside forces.

Many high school students first see the common brown species or less likely the often smaller green hydra in a science class because these animals are inexpensive to buy and easy to culture in the lab, and they can also be collected from clean ponds and small streams. The green varieties get their color from internal, symbiotic algae that supply a second food source. If you collect some filamentous aquatic vegetation and water from a clean pond and allow it to settle undisturbed in a water-filled tray (preferably glass), you may watch the hydra expand to their full length. Some of these may have "baby" hydra developing on their stalks. If you wish to maintain them for longer periods, feed them very early stages (nauplii) of brine shrimp that you hatch (and then rinse) from the fertilized crustacean eggs obtainable from most local pet stores. These fertilized eggs are often labeled by the misleading name "sea monkey" eggs.

Freshwater cnidarians also include two species worldwide of freshwater jellyfish (*Craspedacusta sowerbii*) that are distant relatives of their marine cousins. While those jellyfish (Fig. 2.8) are moderately common in clean ponds, they are rarely seen by humans—even by people who live next to the pond. The three explanations for their mysterious presence are that (1) they are universally very small, about the size of a dime when in their larger and brief medusa (jellyfish-like) life stage; (2) they are mostly translucent; and (3) they live the vast majority of their lives on the pond bottom or on vegetation in a hydra-like form that is even smaller than a normal hydra and only transform for a few days into the floating medusa-like jellyfish. The way people often learn about them is when someone encounters them in a pond and then inquires with a local college professor, which then often leads to an article in the local newspaper. Unlike the often dangerous marine jellyfish, you would be highly unlikely to notice the sting of these small and rarely seen creatures.

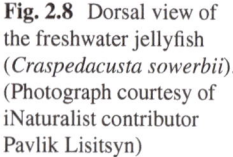

Fig. 2.8 Dorsal view of the freshwater jellyfish (*Craspedacusta sowerbii*). (Photograph courtesy of iNaturalist contributor Pavlik Lisitsyn)

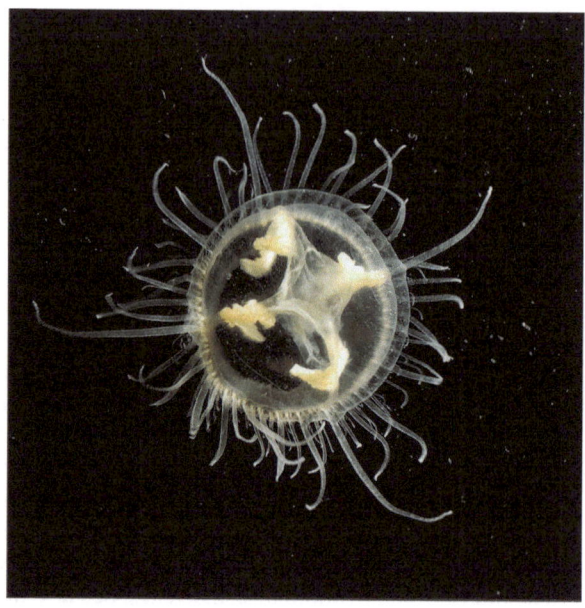

Freshwater Flatworms: Phylum Platyhelminthes

Another common freshwater animal studied living in high school and college biology classes is a flatworm, a member of the mostly marine phylum Platyhelminthes. The most commonly observed specimens seen in classrooms are flattened, ribbon-like individuals called "planaria," which are in the traditional class Turbellaria. This phylum represents a major evolutionary leap over hydra because they have three cellular layers (ectoderm, mesoderm, and endoderm) instead of two. They also have eyespots and a neural area serving as a simple brain. While some organ systems are advanced in this group, planaria still rely on hairlike cilia for locomotion, though some can swim for short distances using a twisting motion of their whole body. Even though their gut has only one opening to the exterior instead of the much more efficient single-pass system characterizing more complex phyla with its separate entrance and exit, many scientists believe this phylum led through multiple stages of evolution to all higher phyla.

Turbellarians like planaria are often studied in high school and college labs because they have a remarkable ability to regenerate. Indeed, you can cut off a planarian's anterior region (often called its "head"), and it will regrow a new one under the right laboratory conditions. Even more remarkable is that you can partially divide it in half longitudinally, and two connected heads may develop!

Freshwater planaria tend to be a dull brown or black, which helps them to hide from predators in their local habitats. In contrast, their marine relatives are often very bright and colorful, but that typically reflects the fact that those marine species had previously ingested stinging cells from corals and then moved them to the flatworm's outer tissues. These foreign cells can be maintained without discharging until a predator tries to attack the marine worm and gets a painful surprise!

To collect freshwater planaria, obtain some decaying beef or chicken, and place it on the bottom of a clean stream inside a trap with a mesh just small enough to hold the bait while also excluding predatory fish and crayfish. After a day or two, you may have planaria attached to the meat, but don't give up too early if you do not first succeed. In addition to eating decaying tissue, planaria will also seek living prey such as hydra, various worms, and tiny crustaceans.

Moss Animals: Phylum Ectoprocta (or Bryozoa)

Ectoprocta, or more commonly called Bryozoa, is a phylum of mostly marine colonial species but which also contains some unusual freshwater representatives. Their odd common name—moss animals—seems to have come from early scientists who confused them initially with plants. They are included in this chapter mostly because you may see their massive colonies in the summer suspended from beneath a dock or less likely (in terms of visibility) from tree roots or branches hanging in a river. They also may form roundish colonies that are free-floating in rivers and lakes. They would likely appear as a large gelatinous mass with numerous small black dots or rosettes, each formed of 12–18 individual microscopic animals. The overall mass can be of various colors including dark gray and even green. The individual colony members filter suspended particles from the water as their primary food source. While most colonies are somewhat innocuous and pose no significant problems to humans, they can also on rare occasions be substantial biofouling animals that may interfere with the functioning of irrigation, water treatment, and industrial cooling systems by clogging the system's water flow pipes.

Mussels and Clams: Phylum Mollusca

The phylum Mollusca is hugely successful in the world's oceans and to a lesser extent in freshwaters. In North America freshwaters, we have the world's largest diversity of native unionid "pearl" mussels (Fig. 2.9) as well as a lower diversity of small, native "pea" or "fingernail" clams. These freshwater mussels occur in relatively clean, small streams to large rivers with sandy to rocky bottoms and in the shallow waters of many lakes, but their distribution within those systems can be very patchy. For food, they filter mostly algae, protozoa, and some bacteria-laden particulate organic matter floating in the water or lying on the stream bottom. Native species in North America, which are primarily in the family Unionidae, occur in most parts of the USA but are concentrated in southeastern and midwestern states. The largest is the "washboard" mussel (*Megalonaias nervosa*), whose bivalved shells can reach 8 in/20.3 cm in breadth! They can easily live for 30 years, and some

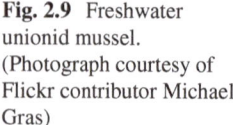

Fig. 2.9 Freshwater unionid mussel. (Photograph courtesy of Flickr contributor Michael Gras)

have reached the century mark in ages past based on a count of their shell's annual growth rings. Native Americans occasionally ate these mussels, but they are purported to be considerably less tasty than the best marine mussels and oysters.

The existence of freshwater mussels worldwide has been severely threatened by pollution, damming of streams, and invasion of competitive mussels and, oddly enough, previously by the production of buttons and marine pearls. In the late 1800s and early 1900s, before the widespread use of plastics, pearly buttons for clothes were made from punched-out sections of unionid shells. This caused a severe decline in numbers of individuals and species across much of their native ranges. A more species-limited problem now occurring is that the shells of some unionid mussels are employed in the production of marine pearls. In both types of bivalved mollusks, the pearls are slowly built up by successive layers of the same material they use to build their shells (aragonite and conchiolin). When the mollusk attempts to reduce irritation from a hard particle which has become lodged by nature or humans between its hard pearly shell and its soft tissue mantle, it covers the particle with successive layers of protective nacre. The natural and especially "cultured pearls" from marine oysters are almost always much more symmetrical and valuable than those of freshwater mussels, but you can find the rarer and still attractive strings of freshwater pearls in some jewelry stores, or you can order them online. Only marine oysters seem to be used to produce cultured pearls, but unfortunately, a market exists for the shells of some freshwater mussels which are sold to foreign companies to grind up and use as the "seed" to produce cultured marine pearls from their oysters! This practice is adding significantly to the decline of the natural diversity and density of some of our native benthic species in the southeastern USA.

A much more significant threat to native species in North America comes from invasive bivalve species. Since the middle of the last century, the Asian or Asiatic clam *Corbicula fluminea* (Fig. 2.10) has been common in many North American rivers, especially those with sandy bottoms. This small clam is also present as an invader in Europe and South America where it competes for suspended food with

Fig. 2.10 Asian freshwater clam (*Corbicula fluminea*). Photograph courtesy (of Flickr contributor "coniferconifer")

Fig. 2.11 (**a**) Clump of the zebra mussel, *Dreissena polymorpha*, on a rock; photograph courtesy of Flickr contributor Sam Stukel; (**b**) the quagga mussel, *D. bugensis*, with siphon extended. (Photograph courtesy of a Flickr contributor from the NOAA Great Lakes Lab)

the native bivalve mussels. While this clam has caused some problems in blocking intake pipes from rivers, their negative effects are far exceeded by two species of European invaders: zebra and quagga mussels (*Dreissena polymorpha* and *D. bugensis*, respectively; Fig. 2.11a, b). These mussels attach extensively to many solid surfaces—including boats, pipes, our native mussel species, and even some crayfish—using their sticky, external byssal collagenous threads. This can lead to the physical displacement or actually smothering of the native species, which prevents them from feeding. Unfortunately, they can also attach to a boat hull or motor and be transported temporarily out of water from one lake to another, thereby spreading to new habitats. Unlike native mussels, *Dreissena* species produce planktonic larvae (= veligers) that float downstream. This reproductive process disperses these species more rapidly than can the technique used by native mussels, as described below.

Bizarre Reproductive Deceptions The reproductive process of native, sedentary, unionid mussels involves releasing into the surrounding waters a larval stage known as glochidia (Fig. 2.12). This is a dangerous life stage for the mussel larva because its survival depends on finding a suitable fish host within a few hours or rarely up to a couple of days. The larva must attach to the fish's gills or external tissue either as a mere haven to hang out during maturation for a few weeks or months or as a

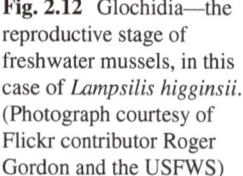

Fig. 2.12 Glochidia—the reproductive stage of freshwater mussels, in this case of *Lampsilis higginsii*. (Photograph courtesy of Flickr contributor Roger Gordon and the USFWS)

temporary "restaurant" home where it can feed on blood until it reaches a size where it can detach from the fish, settle on the stream bottom, and begin collecting nutrients from the water or surrounding bottom. To enhance the chances of a parent's offspring finding a suitable fish host, some mussel species modify the shape of their fleshy mantle—which is the outer tissue between the shell and internal organs and which secretes the shell. In a few species, the mantle is altered to look like a crayfish, minnow, worm, or other prey items in order to attract a predatory fish. When the fish attacks, the mantle bursts and its enclosed larvae are released onto the fish. In another similar approach, some mussels open their bivalved shells to make it appear that they have recently died. When a fish pokes its head between the valves of the shell to eat the supposedly decaying contents, the surrounding shell snaps shut, temporarily holding the fish long enough for the reproductive units (glochidia) to be released and then attach onto the newly acquired fish host. To be honest, I do not know how many fish have fallen for this ruse a second time!

Phylum Arthropoda

The most successful, multicellular phylum by far is Arthropoda, with a large contingent of species inhabiting freshwaters. Included here is a diverse fauna of insects in all aquatic habitats, a relatively few mites, and abundant benthic crustaceans as well as the planktonic cladocerans ("water fleas") and copepods discussed earlier in this chapter. Freshwater mites—while present in most habitats—are generally tiny commensal and ectoparasitic species, commonly living on aquatic insects and benthic crustaceans. You can find them by diligently searching on the exterior of aquatic insects and sometimes on crayfish, but their identification is relatively difficult even for many professional taxonomists. Among the larger freshwater crustaceans are benthic species of crayfish, shrimp, and crabs, as discussed below, but let's first explore a very few of the freshwater insects before closing with a namesake of this book—the fairy shrimp (and other large branchiopods like clam and tadpole shrimps).

Freshwater Insects

While the vast majority of insects are terrestrial for all their lives, insects are still very important components of freshwater benthic communities. Freshwater species usually spend most of their lives in an aquatic life phase before emerging as flying adults for a few hours to several weeks. While whole books have been written on flies alone (order Diptera), I am limiting the coverage here to a few taxa that are particularly unusual in some way that may either appeal to, surprise, or disturb my readers! I have avoided delving too deeply into taxonomy in this book, but it is worth noting that insects (class Hexapoda) are now considered members of the larger subphylum Crustacea (which also contains shrimp, crayfish, barnacles, and many other groups) within the phylum Arthropoda. In addition to the few insect groups discussed below, I hope you will investigate the case-building caddisflies (order Trichoptera) in Chapter 5.

The Dragon and the Damsel Among my favorite aquatic insects are members of the ancient taxonomic order Odonata, which is composed of dragonflies (Anisoptera; Fig. 2.13a, b) and damselflies (Zygoptera; Fig. 2.14). I have occasionally studied and very much like both groups of odonates in their crawling aquatic stages (= nymph or naiad) and when they are flying terrestrial adults. However, my admittedly biased view is that the typically brownish larvae tend to be somewhat ugly, while the adults are commonly quite beautiful! This is in part because the head of a

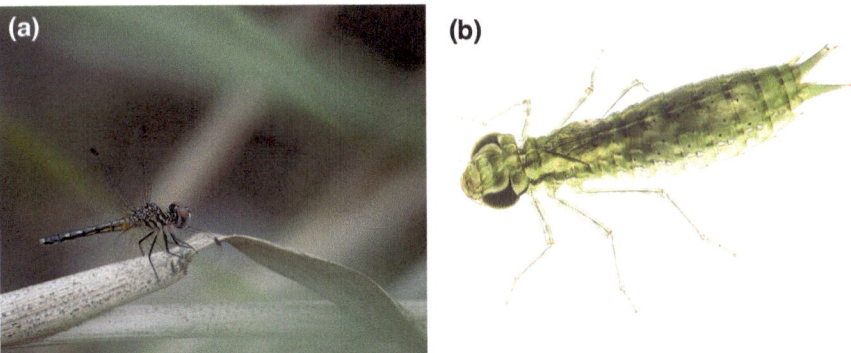

Fig. 2.13 (**a**) Dragonfly adult (Anisoptera); photograph courtesy of Christopher F. Frazier; (**b**) dragonfly nymph. (Photograph courtesy of Sam Stukel (US Fish and Wildlife Service) at Flickr)

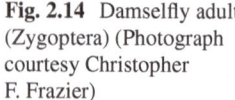

Fig. 2.14 Damselfly adult (Zygoptera) (Photograph courtesy Christopher F. Frazier)

naiad has a mug that only a mother could love, while adults can be strikingly color-ful. The larval life stage is typically the longest and may last months to even a year or more in colder latitudes. Naiads tend to remain motionless on the bottom to avoid predators and possibly capture small prey that pass within reach of the odonate's extendible mouthparts. Adults can be frequently observed around a pond in the spring and summer, with a female periodically dipping her abdomen in the water to release eggs. At other times, the adults can be seen darting around the land in search of prey or resting on nearby vegetation. They can be carefully captured with a but-terfly net and later released (like all freshwater species) where possible. Be careful in handling the adults, however, because their wings are quite fragile.

The odonate life begins when the eggs are laid in a stream, pond, or lake. The aquatic phase includes a species-specific progression of 9–17 molts during which the odonate repeatedly casts off its smaller outer skin for a more suitable and larger set of biotic "clothes" before finally crawling out of the water and perching on a nearby plant stalk to undertake the final molt into a flying adult. This dramatic tran-sition from an extended aquatic larval stage to an often brief adult stage is typical of most aquatic insect taxa except for a few large orders where the adult primarily stays submerged within or upon the aquatic habitat. Nymphs may spend several months to a few years growing and developing, with the length of the period depen-dent on the latitudinal climate. Adults, however, rarely survive more than a month or two and might live for only a few days depending on the species and the abun-dance of potential predators.

While possibly considered ugly, the larval (naiad) head is remarkably functional with its extendible labrum. When retracted, the labrum fits upon the front of its head like a mask. But in action, this structure can be rapidly opened and thrust far for-ward as a living "stalk" to grasp a prey in a deadly toothlike embrace and bring the victim back to the odonate's mouth. Dragonfly and damselfly naiads are superbly efficient predators, consuming many kinds of mostly smaller invertebrates including other smaller odonates of the same or different species. But, the enormous dragonfly naiads in the family Aeshnidae (called aeshnids, darners, and hawkers) can slay and consume piecemeal much larger prey, including tadpoles and minnows. They grasp these larger prey with their legs and gradually eat them alive! In contrast, odonate naiads can easily fall prey to fish and many larger aquatic insects. Consequently, they must spend most of their day hidden upon or partially buried within the surrounding dark sediments (most dragonflies) or clinging to concealing upright vegetation (most damselflies).

Adult odonates are also efficient predators. As strong fliers, they catch most of their prey on the wing. Their prey include many types of insects, such as butterflies, grasshoppers, other odonates, and even small bees, as well as spiders. At the head of the list of enemies for the adult odonates are birds and bats, but frogs eat many of the emerging "teneral" odonates, and nonflying insects, spiders, lizards, and some mammals can capture them, especially when an odonate is resting. Many odonate species in Canada and the northern USA migrate in late summer and early fall toward warmer regions (e.g., Gulf Coast, Mexico, and even the East Indies) only to return to northern regions in the spring.

Here Today, Gone Tomorrow: Mayflies The taxonomic name of mayflies— Ephemeroptera—is very apropos because the life span of adults is typically measured in hours rather than days or weeks. Their common name is not globally accurate, however, because their emergence is not always in May, especially outside the north temperate zone. These mayflies—also called shadflies or fishflies in the USA and up-winged flies in the UK—are unique among insects in having a terrestrial, fully winged, preadult stage (the subimago) in addition to the nymph and adult stages.

In the larval stage (Fig. 2.15a), mayflies depend mostly on a diet of algae and detritus found primarily in clean streams and some lakes and ponds. A very few species prey on larval chironomid midges. Many stoneflies, other predatory insects, and fish eat these larvae, while birds and spiders account for the demise of many adult mayflies. Their consumption by fish is the reason many avid fly anglers note the temporary presence of adult mayflies and then employ mimics to entice a local trout to grab an artificial mayfly lure.

Adult mayflies (Fig. 2.15b) have coordinated emergence as part of the mating. Males often form huge swarms into which females will temporarily enter to mate. During the adult stage, mayflies do not feed—indeed, their mouthparts are only vestigial—though they may drink if they survive more than a day. Females usually fly upstream after mating to lay their eggs so that subsequent larval populations will eventually drift downstream and replenish the downstream losses from abiotic and biotic forces and emergence from the water. Mayflies migrations would probably go largely unnoticed by humans other than by anglers were it not that their coordinated emergence to flying adults can be huge. In fact, the swarms can be so dense as to represent hazards to car drivers. One spectacular emergence event in the USA was even noted on Doppler radar at a local weather station!

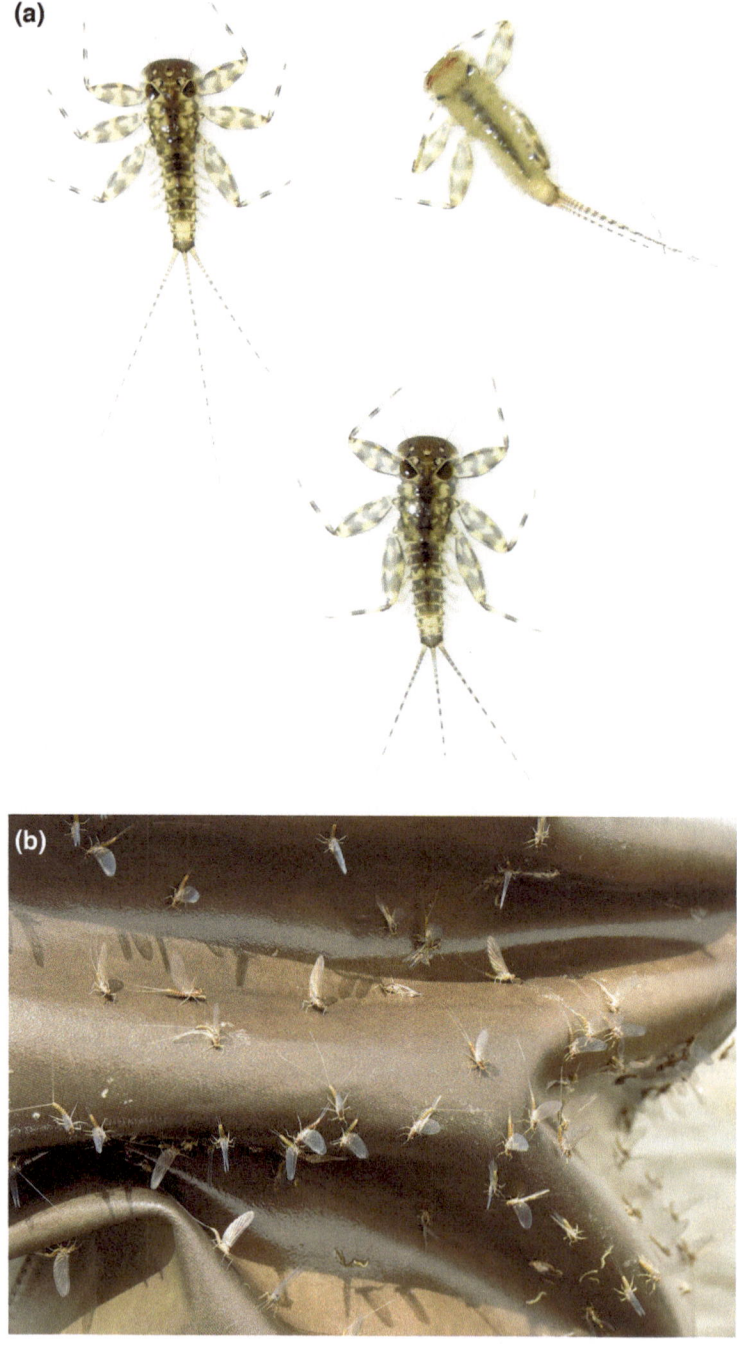

Fig. 2.15 (**a**) Mayfly larvae (Ephemeroptera) and (**b**) a swarm of now resting adult mayflies. (Both photographs are courtesy of Flickr contributor Tom Koerner)

Fig. 2.16 (**a**) A 3 in/7.6-cm-long larval hellgrammite (Megaloptera, *Corydalus cornutus*); photograph courtesy of Flickr contributor Judy Gallagher; (**b**) adult dobsonfly. (Photograph courtesy of Flickr contributors Andy Reago and Chrissy McClarren)

Pinch Like There Is No Tomorrow: Hellgrammites Most aquatic insects cannot impart a significant bite on humans, but watch out if you carelessly pick up an aquatic larva of the insect order Megaloptera (meaning "large wings") with their long, strong, and sharp mandibles. As terrestrial adults, these insects are known as eastern dobsonflies or fishflies (Fig. 2.16a, b) in North America, but they are also present in much of the northern hemisphere as well in South America. They tend to be found in clean, well-oxygenated water.

Larval hellgrammites typically exceed 1 in/2.5 cm in length, but really big ones can reach 2 to almost 3 in (5–7.6 cm). Like many other insects, they have three pairs of thoracic appendages. However, with only a cursory glance from the avid naturalist, the six to eight pairs of lateral abdominal filaments used for breathing may give the false impression of more legs. These insects generally take 1–3 years to mature in streams. They emerge afterward on land where for a while they hide in a terrestrial burrow or other shelter to pupate safely before later transforming into flying adult Dobson flies.

Insects That Bite Us Pretty much wherever you live from the polar regions to the equator, you will often find yourself in warmer seasons trying to avoid the bites of several types of insects in the order Diptera (true flies), all of which spend their larval stages in an aquatic habitat. While the vast majority of dipteran species do not bother humans and many taxa are ecologically important players as aquatic larvae (such as mosquito "wigglers"; Fig. 2.17a), adult mosquitoes (family Culicidae; Fig. 2.17b) are omnipresent and impart bites to most mammals. Many also carry parasitic vectors such as malaria, as well as arboviral diseases including yellow, West Nile, and dengue fevers. Biting midges (family Ceratopogonidae) are called by a variety of common names, including "no-see-ums" because of their small size (usually less than 1/10 inch long). One reason that they are annoying is that you may not know you have been bitten until the next day when you see the cluster of red dots on your skin. In contrast, the more abundant midge larvae of the family Chironomidae (Fig. 2.18a) are extremely important in their aquatic phase, especially as food for other insects, and generally do not bite as adults (Fig. 2.18b), though they may be considered pests when swarming. In contrast, one interesting group of dipterans are called "blood worms" (Fig. 2.21a), not because of their feed-

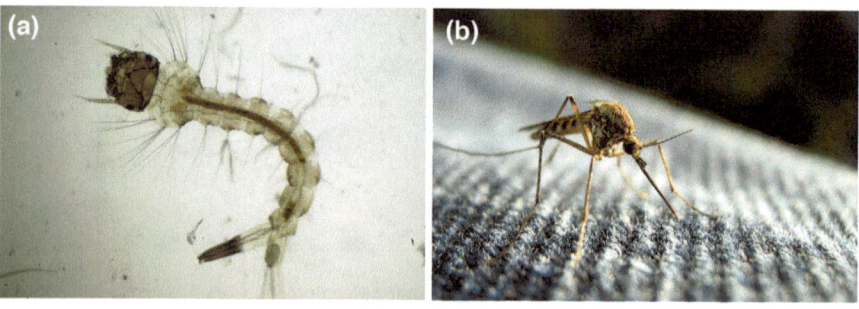

Fig. 2.17 (**a**) Larval mosquito (Culicidae); courtesy of Flickr contributor Rob Cruickshank; (**b**) adult mosquito. (Photograph courtesy of Flickr contributor "turkletom")

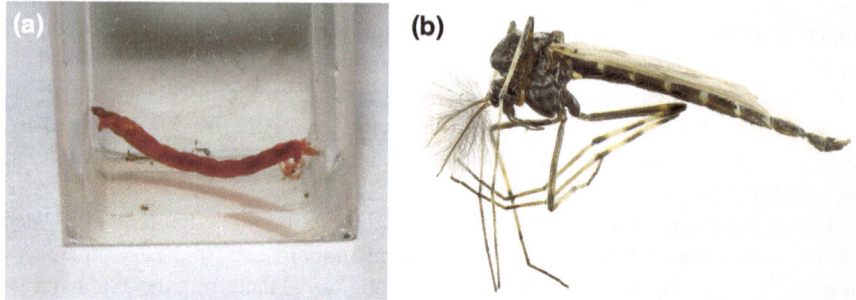

Fig. 2.18 (**a**) Larval midge (*Stictochironomus sticticus*); photograph courtesy of Flickr contributor Rob Cruickshank; (**b**) adult chironomid midge. (Photograph courtesy of Flickr contributor Janet Graham)

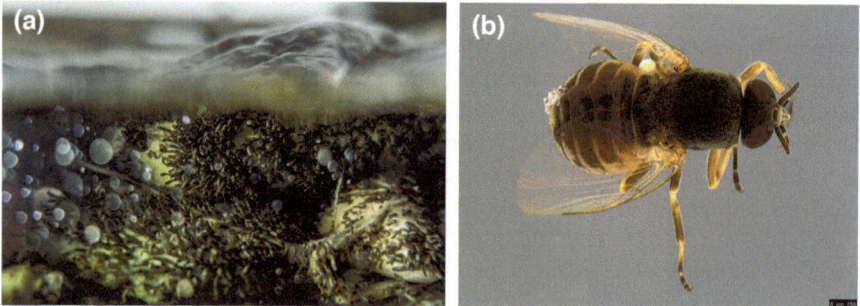

Fig. 2.19 (**a**) Hundreds of larval blackflies (Simuliidae) in the St. Mary River; courtesy of Flickr contributor at the Glacier National Park Service; (**b**) adult blackfly from Australia (photograph courtesy of Flickr contributor Donald Hobern)

ing habits but instead as a reflection of their bright red color from the unusual presence of hemoglobin in their blood. Seeing them is usually a sign that they are living in an aquatic area with low oxygen content. Finally, blackflies (family Simuliidae; Fig. 2.19a, b) are common in Canada and the northern USA. They are larger than biting midges and pack a bigger wallop! You may not know you have been bitten by an adult (Fig. 2.19b) until later when you find a clot of dried blood among your head hair!

Yuck ... Cockroaches! In one sense, it should not surprise you that many cockroaches can tolerate minimal immersion in moist habitats because you may have been shocked by seeing them climb up from or down into a drainpipe. While far less than 1% of cockroach species (order Blattodea) are actually aquatic, there are two groups of semiaquatic roaches. The adults and immature stages (= nymphs) of one species can be found in the small, tropical arboreal pools formed by flowering bromeliad plants, while in the other species, only the nymphs live in lakes, rivers, and small streams (both nymphs and adults may occupy nearby trees). Such species— which pretty much look like the "terrestrial" cockroaches most of us have seen— can live on the water surface and may only submerge to escape predators and search for food.

Hop-a-Long Grasshoppers The grasshopper is another animal you would not expect to find as an intentional resident of a stream, but in the tropics, these semi-aquatic animals are associated with aquatic plants. They pretty much look like terrestrial species, but some sections of their legs are somewhat enlarged to aid in moving the adult through the water. A few can swim to a limit extent, and all feed on aquatic weeds, as well as lichens, mosses, and organic debris. However, all submerge for only brief periods before returning to the air-water surface. When underwater, they use a body structure (plastron) to trap a bubble of air which they take below the surface because they cannot directly extract oxygen from water.

Decapod Crustaceans: Crayfish, Crabs, and Shrimp

The ease by which you can find crustaceans in freshwaters of all continents other than Antarctica will not come as a surprise to many readers, but you may be mildly astonished to learn that the type commonly encountered in the wild greatly depends on your home continent. While most crustaceans living in surface streams and lakes on all continents are tiny copepods and "water fleas" (Cladocera; see earlier discussion), the species best known to the public are much larger animals in the order Decapoda (meaning ten feet or legs, though some of these appendages are often adapted for other purposes). Within Europe and North America, decapods usually consist of three types: (a) the abundant crayfish (very diverse in the USA—especially in the southeastern states—with a few native species still remaining in Europe); (b) less commonly observed freshwater shrimps, including species like *Palaemon kadiakensis* and *Macrobrachium ohione* (Fig. 2.20); and (c) estuarine crabs that move short distances upstream into freshwater rivers. However, on a global basis and especially south of the Tropic of Cancer, it is multiple species of freshwater crabs and shrimps that greatly dominate the decapod fauna in streams and lakes. In the USA, crayfish are the most widely consumed, totally freshwater

Fig. 2.20 River shrimp (*Macrobrachium ohione*). (Photograph courtesy of Flickr contributor Alex Solis)

Fig. 2.21 The shrimp *Macrobrachium rosenbergii*, a resident at various times of marine, estuarine, and freshwaters. (Photograph courtesy of Flickr contributor Jonathan Vera Caripe)

crustacean. In contrast, the shrimp *Macrobrachium rosenbergii* (Fig. 2.21)—a marine-estuarine-freshwater species native to Southeast Asia—is the supreme species worldwide in culinary markets. While this species has also been imported and is now raised in North America, the shrimp market in the USA is still dominated by multiple species of shrimp (e.g., *Penaeus* sp.) living in coastal marine-estuarine habitats.

All crustaceans have a relatively hard, external carapace which provides bodily support and some protection from predators. This exoskeleton must be shed periodically to allow growth and development, while a soft, underlying new exoskeleton then replaces it, expands, and hardens. During the hardening process, which takes place over the next day, the animal is very susceptible to predators.

Crayfish Crayfish—which are also known by other names in North America, such as crawdads, crawfish, and even mudbugs—have been consumed for centuries on this continent, especially in Louisiana and other southeastern states where they are

Fig. 2.22 Front view of
the red swamp crayfish
Procambarus clarkii in
defensive posture.
(Photograph courtesy of
Flickr contributor Javier
Colmenero)

most abundant and diverse. They have also been part of the diets of many indigenous peoples in North America for even longer periods. Starting late in the twentieth
century, they even appeared as "gourmet items" on some restaurant menus in
New York and other large northern cities. At first the species most frequently available in those restaurants was the red swamp crayfish, *Procambarus clarkii*
(Fig. 2.22). This and other crayfish are very common in small streams, ponds, and
wetlands, especially where large fish are absent or less frequently encountered.
Crayfish may live in burrows on land near an aquatic habitat (see Chapter 5) or
thrive in the stream or lentic system where they can find protection in the bank or
under rocks, wood, or aquatic grasses from predatory fish, birds, and mammals. All
are omnivores and will become predators and even cannibals when the opportunity
arises. Like other crustaceans, they periodically shed their "skin" and replace it with
the underlying version. At this life stage, they are extremely vulnerable to predation
until the new exoskeleton hardens.

After copulating with a male crayfish, the female stores her fertilized eggs in a
brood pouch, which she may carry for months. Therefore—and in contrast to marine
lobsters—the mating process in freshwater crayfish is not linked in time to ecdysis
(molting). Eventually she attaches the fertilized eggs to her abdominal appendages
and periodically aerates the eggs by moving her tail. Once the young hatch, they
remain on her abdomen as they develop over the next several weeks. As they grow

larger, they periodically leave for brief periods before scurrying back to her protective tail when danger threatens. Eventually, they permanently leave the mother's care to explore the dangerous world where being eaten by one of many predators is a high probability.

Unfortunately from an ecological perspective, American crayfishes, especially *Procambarus clarkii*, have been exported live for aquaculture to other US states as well as to Europe and other continents. In most cases, introduction of this invasive species to different continents has come at the expense of native crayfish and other invertebrate species. However, in Africa, their presence—while generally negative for native crustaceans as well as some other animal and aquatic plant species—has also had a positive effect from reduction of the trematode parasites that infect humans. This results from the crayfish consuming the trematode's intermediate snail host. Some efforts have also been made to import the much larger redclaw crayfish (*Cherax quadricarinatus*) and yabbies (*Cherax destructor*; Fig. 2.23) from Australia into both the western hemisphere and countries of Southeast Asia. These crustaceans reach the size of small lobsters (making them attractive for restaurants), but they could severely damage populations of native freshwater unionid mussels and crayfishes in the wild if they reach the southeastern USA.

Freshwater Crabs and Shrimps   In tropical and subtropical regions of the world, a very diverse and abundant fauna of freshwater crabs and shrimps dwarf in numbers those of resident crayfish, but they are generally not harvested for public consumption. The exception is for a few species, such as the shrimp *Macrobrachium* (Fig. 2.21), which can grow fairly large and which are then raised in aquaculture ponds around the world, including in coastal USA. Unfortunately, this aquaculture process in Asia often involves cutting down large areas of coastal mangrove forests, resulting in massive soil runoff that can easily smother any nearby, offshore coral reefs. These Asian ponds usually produce only short-term profits (perhaps a decade at most) before they are abandoned, leaving ecological devastation in their wake to the once thriving mangrove forest ecosystems and other estuarine habitats. This tendency to ignore or trade ecologically significant damage for short-term monetary profits is a process that has become an all too common theme globally.

Fig. 2.23 A gravid Australian yabby (*Cherax destructor*). (Photograph courtesy of Museum Victoria's "Catching the Eye")

While insects are generally the most important invertebrate of temperate zone streams, freshwater crabs and shrimps are often the major components of tropical streams. Crabs are opportunistic predators to some extent (e.g., feeding heavily on snails in Africa's Lake Tanganyika), but they are primarily nocturnal scavengers that shred and consume leaf litter and other coarse organic materials as the main macro-decomposers in tropical streams. Tropical freshwater shrimp species are often more abundant in medium to higher elevation stream sites lacking medium-to-large fish. Like crabs, they also consume detritus and benthic algae but are also opportunistic scavengers. A high proportion of freshwater crab species are threatened with extinction, including, for example, 98% of the 50 species currently residing in Sri Lanka.

A much less common group of freshwater crustaceans are members of the decapod infraorder Anomura and the family Aeglidae, with fewer than 90 species existing worldwide. All are very closely related to marine hermit crabs, but neither group are in the same taxonomic grouping as true crabs (Brachyura). All are restricted to freshwater rivers, lakes, and caves of the southern portion of South America (Fig. 2.24), and at least a third are considered threatened. A strange thing about these crustaceans is that even given their very close relationship to marine hermit crabs, they now survive without the protective homes of dead gastropod snails that characterize all marine species in this infraorder. Only one anomuran crab species retains a shell in freshwater habitats (*Clibanarius fonticola*), and it is restricted to the southwestern Pacific island of Vanuatu.

The giant coconut crabs (*Birgus latro*; family Coenobitidae; Fig. 2.25) are in a very unique group of anomuran crabs whose adults are shell-less. These "robber" or "thief" crabs live on land and in trees as adults, but they release their larvae into local streams or directly into the ocean for their offsprings' initial development, which includes living in snail shells until they move on land. Although they really do not deserve coverage in a freshwater book, I confess that I am mentioning them merely because they are "cool critters"! If you are lucky enough to see one of these huge crabs on an island in the Pacific or Indian Ocean, be sure to keep your fingers

Fig. 2.24 A shell-less, anomuran crab (*Aegla papudo*). (Photograph courtesy of iNaturalist contributor Asiel Olivares)

Fig. 2.25 Giant coconut crab (*Birgus latro*) from Diego Garcia in the Chagos Archipelago. (Photograph courtesy of Flickr contributor Drew Avery)

away from them, as their massive claws can rapidly snip off one or more of your appendages!

Life in Ephemeral Environments: Fairy Shrimp and Both Near and Distant Relatives

As their name implies, ephemeral wetland pools are aquatic systems that only temporarily hold surface water. The wet period may be as short as a week or two or as long as multiple years. Ephemeral wetlands include systems that refill multiple times per year and others that may not fill again for years or even decades, depending on weather patterns. A critical biological characteristic is that fish rarely survive in these wetlands, thereby allowing invertebrates to live there that would otherwise be rapidly consumed.

Animals have basically three options for life in temporary wetlands, other than just visiting these pools for a drink of water and a meal of resident invertebrates. One strategy is to hatch, grow rapidly, reproduce within 1–2 weeks in some species, and then produce resistant stages (= fertilized eggs, or "cysts"), thereby allowing the species to survive if the water disappears for months to decades. Examples of

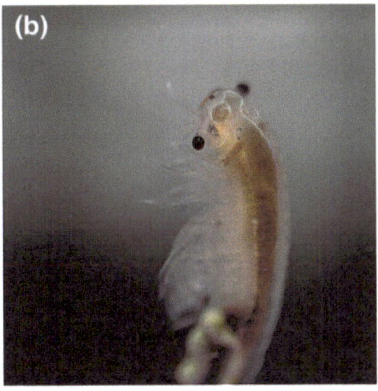

Fig. 2.26 (**a**) Image of a fairy shrimp (*Thamnocephalus platyurus*); (**b**) photograph of a live fairy shrimp (*Branchinecta* sp.). (Photographs courtesy of Christopher F. Frazier)

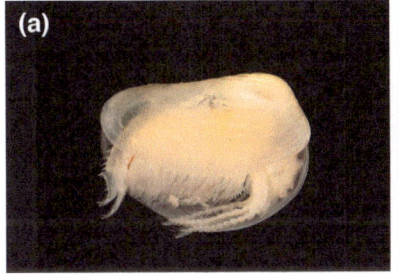
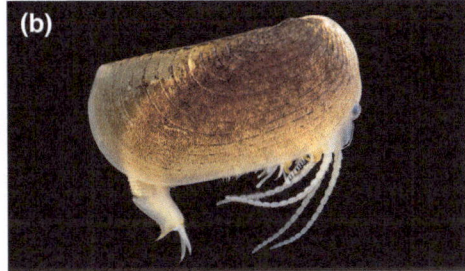

Fig. 2.27 (**a**) Clam shrimp (*Eocyzicus diguetid*); (**b**) photograph of the "playa clam shrimp" (*Leptestheria compleximanus*). (Both photographs are the courtesy of Christopher F. Frazier)

animals exhibiting this ephemeral strategy are large branchiopods (Figs. 2.26, 2.27 and 2.28), cladocerans (Fig. 2.5), copepods (Fig. 2.4), and a few snails, along with various microscopic invertebrates like rotifers (Fig. 2.3) and the ever-present single-celled protozoa. A second strategy is to hatch, grow into adults, depart the pool for short-to-long periods, and then return to lay eggs for the next generation. Among animals exhibiting this second strategy are salamanders (Fig. 2.29), frogs, and spadefoot toads (Fig. 2.30)—some of which are included on state and/or federal lists of threatened or endangered species. And third, they can fly into the wetlands, rest and consume energy from local prey for short-to-long periods, and then fly to other wetlands on their migratory path, which, in one sense, is much like all of us might do on a cross-country driving or bicycle trek when visiting hotels, restaurants, and gas pumps or electrical charging stations! Consequently, the local fauna varies significantly from pools in the moist forested states of the eastern USA to species occupying the more open and ephemeral pools of the typically drier western states.

Among the few residents of ephemeral wetlands that may be recognized by the public are fairy shrimps (order Anostraca). These crustaceans are better known to some kids and adults as the "sea monkeys" that kids can hatch from tiny fertilized

Fig. 2.28 Tadpole shrimp
(*Triops* sp.). (Photograph
courtesy of Christopher
F. Frazier)

Fig. 2.29 Dusky salamander (*Desmognathus fuscus*) crawling through moss. (Photograph courtesy of Flickr contributor Dave Huth)

Fig. 2.30 Plains spadefoot toad (*Spea bombifrons*). (Photograph courtesy of Flickr contributor Sam Stukel (USFS))

eggs (= cysts) purchased in a local pet store. These shrimp are fun to observe in their own right, but they can also be raised as food for some other animals including as a tasty item for aquarium fishes. I also fed young fairy shrimp (nauplii) to green hydra as part of my master's thesis research many decades ago! The label in the store often refers to the species as being *Artemia* (Fig. 2.31; usually *A. salina* or *A. franciscana*), which in the USA occur in the Great Salt Lake of Utah and many other fish-less wetlands and lakes, many of which are somewhat very salty. To scientists, the fairy shrimp are known as members of the taxonomic class Branchiopoda of the subphylum Crustacea. This is a huge group of arthropods that also includes crayfish, crabs, true shrimps, barnacles, insects, and many other taxa. Closely related are clam shrimps (orders Laevicaudata and Spinicaudata; Fig. 2.27) which are enclosed in an organic shell and are more consistently bottom dwellers than fairy shrimp and the often larger tadpole shrimps (order Notostraca; Fig. 2.28), which remind me very superficially of marine horseshoe crabs. Clam shrimps could initially be confused with either ostracod crustaceans or very small, freshwater pea clams. Tadpole shrimps are much less diverse than fairy shrimp, but they are larger and mostly stay closer to the bottom of the wetlands. The feeding behavior of tadpole shrimp stirs up the bottom sediments (bioturbation), thereby increasing turbidity and reducing growth of algae and various macrophytes. It is not surprising that these micro-brutes are the primary predators among the group of large branchiopods in wetlands. They

Fig. 2.31 Mono Lake brine shrimp (*Artemia monica*). (Photograph courtesy of iNaturalist contributor rlawrenz)

forage on fairy and clam shrimps, other small invertebrates, and even each other, and they can be pests foraging in rice fields. Fairy shrimp are much more diverse and widespread than clam and tadpole shrimp. Although you can collect fairy shrimp close to the surface of a playa (= a shallow, often circular wetland that is very abundant in the central and western USA) or other wetland, they normally stay near the bottom where they filter feed on dead organic matter, algae, and smaller invertebrates. Clam and tadpole shrimps feast upon much of the same diet as fairy shrimp, but they almost always do so closer to the sediments. Because all these shrimps need to hatch, grow rapidly, find a mate, and produce fertilized eggs in as little as a full week before the wetland dries, they require a lot of food. In fact, they can consume as much as 40% of their total body mass in as little as a day!

A significant problem for fairy shrimp and their relatives is what to do when their ephemeral wetland dries up! Most will produce environmentally resistant, fertilized eggs (cysts) and stay at least temporarily within the now-dried wetland sediments of their home pool until their home pool again fills and the new generation hatches. Other eggs will be dispersed inadvertently via three main pathways: either carried by the wind or by hitching a ride on (feet, legs, and feathers) or inside (gut passage) migrating waterfowl. Transport inside a bird might seem a bit risky, but scientists have shown that a small number of these eggs can survive passage in the gut until the bird reaches another wetted pool where the avian transport agent "poops out" the eggs or they wash off its outer body. Wind dispersal, which potentially reaches more wetlands, is clearly even more risky because the eggs are much more likely to settle on land surfaces unconnected with wetlands, and thus the enclosed shrimp must wait an indeterminate amount of time to hitch another flight on the wind to an ephemeral wetland. Even if an individual branchiopod makes it to a suitable habitat, its future reproductive success depends on another member of that species of the opposite gender being in that pool simultaneously.

Fairy shrimp survival depends on at least three primary factors. First, they must reach a potentially appropriate wetland in terms of surface area and depth through the semi-random distribution of fertilized eggs via "wind and wing" and meet a potential mate there. Second, the wetland must have sufficient amount of water to allow the shrimp to hatch, grow, and reproduce before the water evaporates. And third, the salt and other ionic concentrations, temperatures, and oxygen conditions must be appropriate, though these shrimps are remarkably tolerant of extremes in both the latter conditions. To better ensure survival of the species, a certain amount of "bet-hedging" is required (much like putting money on multiple numbers in roulette). Of all the eggs of a single species that reach the dry bed of a playa or other wetland, some will hatch soon after it rains, while others will require a period of cold weather and even freezing. And, of course, brine shrimp "sea monkey" eggs might also end up hatching in the aquarium of someone reading this chapter!

For More Information

For more detailed information on freshwater invertebrates, consult the advanced book series *Thorp and Covich's Freshwater Invertebrates* published by Academic Press, Elsevier. The fourth edition contains, as of January 2024, five published volumes edited by multiple scientists (see "Further Readings" below) with information from three continents (Europe, North America, and South America). Another three volumes should be published in the next 3 years for fauna of Africa, Australia/New Zealand, and Southeast Asia and Subtropical China. For a less detailed key for North American waters, see Thorp and Rogers 2011. In addition, a separate volume (Maasri and Thorp, eds. 2023) is available for the Mediterranean fauna.

Further Reading

Maasri, A. and J.H. Thorp (eds.). 2023. Identification and Ecology of Freshwater Arthropods in the Mediterranean Basin. Elsevier.

Thorp, J.H. and D.C. Rogers. 2011. Field guide to freshwater invertebrates of North America. Elsevier, Boston, MA. 274 pages.

Book Series (only the 4th edition is listed below as of January 2024): Thorp, J.H. and A.P. Covich (eds.). Ecology and classification of North American freshwater invertebrates. Academic Press, Elsevier. Books currently published in the fourth edition:

Damborenea, C., D.C. Rogers, and J.H. Thorp (eds.). 2020. Vol. V: Keys to Neotropical and Antarctic Fauna. 1017 p.

Hamada, N., J.H. Thorp, and D.C. Rogers. 2018. Vol. III: Keys to Neotropical Hexapoda. 811 p.

Rogers, D.C. and J.H. Thorp (eds.) 2019. Vol. IV: Keys to Palaearctic Fauna. 920 p.

Thorp, J.H. and D.C. Rogers (eds.). 2015. Vol. I: Ecology and general biology. 1091 p.

Thorp, J.H. and D.C. Rogers (eds.) 2016. Vol. II: Keys to Nearctic fauna. 740 p.

Chapter 3
Something Is Fishy About This Chapter

Although freshwater fishes are discussed in several chapters of this book, it occurred to me that these common aquatic vertebrates deserve at least a relatively short chapter focusing solely on them to go along with coverage in later chapters. For extended coverage of very large fishes and those that can generate electricity, see Chap. 4. For information on marine sharks and rays that have invaded freshwaters, consult Chap. 6. The chapter has a slight bias favoring discussion of the ~800 freshwater species living in North America (out of more than 10,000 freshwater species worldwide), but information on fish inhabiting other climatic zones receive some coverage.

The Origin of Freshwater Fishes

Bony fish species as well as the cartilaginous sharks and rays reach their greatest species diversity in the world's oceans, possibly in part because of the greater habitat diversity and stability there compared to freshwaters. A natural assumption, therefore, might be that bony fishes originated in the world's oceans and later immigrated into freshwater streams and lakes. Actually, the opposite evolutionary pathway is probably true. Indeed, the blood salinity of freshwater fish better matches the very low ionic state of their surrounding waters than does the comparable blood of marine bony fishes in their higher ionic-strength ocean habitats. From an osmotic perspective, freshwater bony fishes are constantly losing salts from their blood and gaining water but at a low rate, whereas the opposite is true for marine bony fishes. This contrasts with sharks and rays whose diversity has always been strongly centered in marine habitats. Their osmotic balance is maintained not by salts but instead by urea within their blood, including those few species of sharks and rays that immigrated much later into freshwaters.

Modern bony fishes seem to have evolved perhaps 200–250 million years ago (myr) in the late Paleozoic to early Mesozoic eras with perhaps half of the currently

© The Author(s), under exclusive license to Springer Nature Switzerland AG 2024   43
J. H. Thorp, *The Otter and the Fairy Shrimp*,
https://doi.org/10.1007/978-3-031-64029-2_3

recognized fish orders present by 65 myr. The original habitat of freshwater fish was probably streams, especially considering that lakes are usually younger from an evolutionary perspective, even though a few lakes are extremely old (e.g., Lake Baikal in Siberia is thought to be 25–30 million years old) and contain high species diversity of freshwater fish. In most cases, those ancient lakes have or had a river connection to the ocean and currently receive water from multiple rivers. For example, Lake Baikal in Siberia is fed by many rivers, including primarily the Selenge River of northern Mongolia. Its primary outlet to the Arctic Ocean is the Angara River. Even the Laurentian Great Lakes of Canada and the USA are thought to have been developed their current configuration at the end of the last glacial period around 10–12,000 years ago, and their waters reach the Atlantic Ocean via the St. Lawrence River.

Freshwater Habitat Requirements

Habitat choice varies among species and reflects some combination of the habitat's permanence, temperature, oxygen content, chemistry, physical habitat structure, food availability, and the type/abundance of predators and competitors. In general, the diversity of fishes tends to increase from upstream to downstream on the path to the sea. And, although you can find small species throughout rivers, the presence of large-bodied species increases downstream. Moreover, larger rivers tend to have more habitat complexity in the lateral riverscape where water currents are lower (thus less stressful flow), habitat diversity for fishes increases, and both prey diversity and numbers tend to rise.

Habitat permanence and existing conditions as well as diversity of local aquatic habitats all affect most fish species. Only a few fish can tolerate drying of their aquatic habitat by means other than dispersal to new sites. An exception is the African lungfish (*Protopterus annectens*; Fig. 3.1) which can form a protective cocoon in the river bank and breathe with its lunglike organs.

As a general rule, freshwater fishes occupy habitats that are neither too hot (most exist in habitats considerably under 90°F/32°C) nor too cold (above freezing from top to bottom) and that have sufficient dissolved oxygen. Warmer habitats come with the additional disadvantage that oxygen content decreases with increasing temperatures above freezing. Massive die-offs can occur during heat waves, a factor most likely to increase with ongoing climate change. Inadequate oxygen conditions can also occur from decomposition of organic matter and from oxygen uptake by dense populations of cyanobacteria and other autotrophic organisms at night when photosynthesis ceases. Oxygen levels in unpolluted waters increase with stream turbulence, as in rocky headwaters.

Species vary considerably in their habitat preferences, but in general very few freshwater species tolerate a level of salinity comparable even to the levels found downstream in the upper part of estuaries (0.5–5 parts per thousand salt) much less in full-strength seawater. Even mildly saline conditions within inland ponds and

Fig. 3.1 African lungfish (*Protopterus aethiopicus*). (Photograph courtesy of Flickr contributor Joel Abroad)

lakes are prohibitive to fishes and most invertebrates (exceptions include fairy shrimp and brine flies). Conditions are further aggravated by the natural presence in the watershed of toxic metals and metalloids, such as arsenic, antimony, and copper. Some notable exceptions to these saline barriers are found among species that migrate upstream from the ocean (= anadromous) or downstream to the sea (= catadromous), including fishes like eels, lampreys, salmon, shad, sturgeon, and steelhead trout. These taxa migrate between freshwaters and the ocean either annually or during a single reproductive life cycle. For example, adult North American eels (*Anguilla rostrata*; Fig. 3.2) reproduce among masses of floating, marine brown algae (*Sargassum* kelp) in the ocean, and the resulting tiny catadromous eels—also called glass eels—later migrate as yearlings upstream into freshwaters. Over a decade or more, these American eels grow in freshwater streams from "elvers" to a length of 24–31 in/60–80 cm by feeding on crustaceans, insects, annelid worms, etc., and then they return to the ocean as developing adults. An example of the opposite life cycle is represented by some anadromous lamprey species which live about 4 years in the ocean as predators before migrating into freshwaters where they spawn and soon afterward die in the same stream habitat. Their filter-feeding young that had hatched in freshwaters then grow to reproductive size and eventually migrate back to the sea. Other lamprey species spend their entire lives in freshwater where the adults feed on the blood and flesh of other fish species (Fig. 3.3). Perhaps

Fig. 3.2 North American eel (*Anguilla rostrata*). (Photograph courtesy of Flickr contributor Sam Stukel at the US Fish and Wildlife Service, Mountain-Prairie Lab, photographed at the Gavins Point National Fish Hatchery)

Fig. 3.3 Two adult sea lampreys parasitic on the exterior of a freshwater fish. (Photograph courtesy of Flickr contributor at the NOAA Great Lakes Environmental Research Laboratory)

the best known group of migratory fishes are salmon. Many anadromous salmonid species are hatched from shallow, temporary nests in rivers and grow to juvenile size. They then migrate downstream to the oceans, grow to adults, and live for an additional 3–13 years (depending on the species) before migrating back upstream to spawn a new generation before dying, sometimes in the clutches of a grizzly bear!

Fig. 3.4 Rainbow trout (*Oncorhynchus mykiss*). (Photograph courtesy of Flickr contributor Sam Stukel at the US Fish and Wildlife Service, Mountain-Prairie Lab, photographed at the Gavins Point National Fish Hatchery)

Fig. 3.5 Rainbow darter (*Etheostoma caeruleum*). (Photograph courtesy of Flickr contributor Yankech gary)

Habitat preferences differ dramatically among fish species and reflect their surrounding substrates, diet, size, and sometimes predator presence. Rocky, turbulent streams select for species that can tolerate areas of high flow rates and associated turbulence. The best known to the average citizen are probably various species of salmonids, like rainbow trout (*Oncorhynchus mykiss*; Fig. 3.4), which are favored by many anglers—not to mention hungry restaurant patrons! Less well-known are miniature species in shallow, turbulent headwaters, such as flamboyant rainbow darters (*Etheostoma caeruleum*; Fig. 3.5) and uniformly dark madtoms (e.g., *Noturus flavus*; Fig. 3.6). By the way, the latter is a type of catfish and must be handled very carefully despite its small size because spines in their pectoral fins can puncture the unwary handler and inject a mild venom comparable to the sting of a wasp or honeybee. Diet preferences also affect habitat choice, as different environments vary in prevalent food type and size as well as in the exposure of the fish to predators. Habitat preference is also influenced by life stage. For example, while larval species may swim in schools for some protection, adults are more likely to swim alone or in small groups. The occupied habitat area in ponds frequented by small fish (e.g., mosquitofish) often expands into deeper waters when large predators are rare.

Fig. 3.6 Madtom catfish (*Noturus flavus*). (Photograph courtesy of iNaturalist contributor Reuven Martin)

Fig. 3.7 Snakehead catfish (*Channa* sp.). (Photograph courtesy of Flickr contributor Amaury Laporte)

With very few exceptions worldwide, freshwater fish cannot exist long out of water. Aside from species like the African lungfish (Fig. 3.1) which can survive in an underground cocoon for up to a year, several fish species invasive in North America are noted for their abilities to crawl short distances from one water body to the next. These include two species from Asia and/or Africa: walking catfishes (*Clarias batrachus*) and snakeheads (*Channa* species; Fig. 3.7), both of which use their pectoral fins for stability and then wiggle across the terrestrial landscape from one aquatic habitat to another on humid evenings.

Food density and type also substantially influence habitat choice. Most open-water, pelagic lake fish focus on collecting larger zooplankton in open-water habitat, unless their prey live on littoral zone vegetation where different foraging techniques are employed. Some fish species collect zooplankton by sucking in individual prey, while other taxa swim through the water while continuously sieving smaller prey. River fish specializing on benthic invertebrates tend to forage in oxygenated areas under stream banks, among woody and other vegetation, and in slowly moving water. Lake species that depend on a diet of abundant benthic species such as aquatic insects and worms tend to focus on prey found in the littoral zone among vegetation or rocks. Frugivorous fishes—which are mostly but not exclusively limited to tropical latitudes—alter their habitat choices and diet depending on which fruits are seasonally and locally available and which can be obtained when the fruit falls into the nearby aquatic habitat. Nearly 300 species of fish have been identified as eating fruits at some point in their lives. The continual presence of one type of fruit or another over the year in the tropics allows this specialization on fruits that is not possible in the vast majority of temperate zone streams. One of the more interesting types of habitat choice occurs in fish that feed on terrestrial insects, like the nine to ten fresh and brackish water species of archerfish (genus *Toxotes*; Fig. 3.8). These tropical fish have unusually good eyesight for a fish and can hit insects and other prey up to 10 ft/3 m above the water by "spitting" water at the prey—which they usually hit the first time!

Fig. 3.8 Archer fish (*Toxotes*). (Photograph courtesy of Flickr contributor Joseph Bylund)

Trophic Positions, Feeding Techniques, and Predators

While freshwater fishes are less diverse than their marine relatives, they are vital constituents of almost all inland aquatic systems other than ephemeral wetlands, saline lakes, and thermal streams. All occupy trophic positions above primary producers in freshwater streams and lakes, but the majority can be characterized as piscivores (fish predators), planktivores (primarily consuming crustacean zooplankton), or benthic grazers (eating mostly invertebrates that live on the bottom or on littoral plants). Lepidophagy—a specialist feeding technique where the predator strikes another fish to capture a mouthful of scales and some flesh—is common among many tropical cichlid species and even some piranhas but is much less common in the temperate zone. Algae-eating fish are less diverse than those preying on invertebrates and other fishes, but the common omnivorous carp (*Cyprinus carpio*) introduced to North American centuries ago by Europeans is known to eat bottom algae as well as vascular aquatic plants and invertebrates (newer species were unfortunately introduced last century and escaped confinement). Invertebrates are often their principal prey in open-water and littoral habitats. Relatively few species eat algae or benthic detritus. In larger fish species, there is an increasing focus on consumption of other fish and non-fish prey, including on rare occasions frogs and both swimming rodents and birds.

Fish capture individual prey using one of three primary techniques: (1) prey manipulation (e.g., picking an individual prey from the bottom or off a plant), (2) ram capture from behind a prey by outswimming it and then swallowing it with or without macerating the prey, and (3) suction. The former includes having the jaws bite individual prey, gripping/extracting prey from surfaces, and rasping off flesh (as in lampreys). The last method involves creating a negative pressure gradient so that the small prey are sucked into the mouth of the predatory fish. Nearly half of all bony fish can protrude their jaws, typically with the lower jaw dropping down and forward. This "jaw protrusion" has several advantages, including (a) providing a greater capture area, (b) moving the jaws farther forward (especially useful in grabbing individual zooplankton), and (c) increasing suction capacity.

While it would be understandable to assume that the size of consumed prey would be proportional to the size of the predatory fish, that is not always the case. While the maximum mouth gape and the relative size of predator and prey in freshwaters limit the upper size of the fish's common prey, the relationship is not always proportional. For example, the really ancient and moderately long-lived (averaging 30 years) American paddlefish (*Polyodon spathula*; Fig. 3.9) feeds on zooplankton from this fish's larval through adult stages despite the paddlefish's large size (reaching 4.9 ft/1.5 m long and weighing 60 lb/27 kg). Some sturgeons (family Acipenseridae; Fig. 3.10) are known to reach over 25 ft/7.6 m long (7–10 ft/2.1–2 m being more common now days) and prey primarily on benthic invertebrates and some small fish. The abundant common carp (*Cyprinus carpio*; Fig. 3.11), which have been known to reach 75 lb/34 kg (most are less than 10 lb/4.5 kg), are omnivores that favor a diet of aquatic plants and benthic invertebrates.

Fig. 3.9 American paddlefish (*Polyodon spathula*). (Photograph courtesy of Flickr contributor Sam Stukel at the US Fish and Wildlife Service)

Fig. 3.10 Lake sturgeon (*Acipenser fulvescens*). (Photograph courtesy of Facebook contributor Gary Michaud)

Fig. 3.11 Common carp (*Cyprinus carpio*). (Photograph courtesy of Flickr contributor Sam Stukel at the US Fish and Wildlife Service)

Fig. 3.12 Migration of Chinook salmon (*Oncorhynchus tshawytscha*) in the Red Hill Creek salmon run. (Photograph courtesy of Flickr contributor Michael Hunter)

While the most significant predators of fish are other fish (e.g., schools of piranha), they are also subject to attack by larger vertebrates. Predator types include birds (e.g., ducks, eagles, ospreys, and cormorants), alligators, otters, and bears, with the last group being especially voracious during salmonid migrations (Fig. 3.12).

Fish Reproduction

Fish reproduction varies considerably between marine and freshwater species, with most of the former releasing eggs and spermatozoa into the water, typically in mass spawning. Freshwater bottom dwellers, like catfish, build a nest in some protected area such as among submerged rocks, aquatic weeds, or tree limbs that have fallen into a lake or stream. After the eggs are deposited and fertilized, the males tend to guard the nest until the young hatch and later swim away. Some largemouth bass also build nests but may do so near shore where you can often observe them in relatively undisturbed locations during the early summer. Once an admiring female enters the nests and deposits her eggs (several thousands to tens of thousands), the resident male often must quickly fertilize them and prevent a smaller "sneaker male" from rushing into the nest to add his sperm. As indicated earlier, many species of salmon swim far distances upstream from the ocean, where they scoop out a shallow nest in a fast moving stream in the hope of attracting a female to lay her eggs there for him to fertilize.

Conservation of Freshwater Fish Fauna

When it comes to preservation of freshwater fish fauna, "We have met the enemy and he is us!" (adapted from a Pogo cartoon in 1970 by Walt Kelley commenting on the surrounding pollution in his swamp forest). I say this because so much of our headwater streams through large rivers have been destroyed or severely altered by humans via dams, levees (except for less damaging setback levees), watershed disturbance, chemical and thermal pollution, and foreign species introduced for their desirability by anglers (or by accident). In the defense of the USA, I have to admit that other first-world countries have certainly done similar bad things to their aquatic systems (though species introduction is rarer outside of North America, Europe, and many first-world Asian countries). There are still a few countries remaining with mostly native fauna, including Mongolia which at least between 2010 and 2020 had only one documented invasive fish species. What can we do about this problem? Some good suggestions are to (a) clean our rivers; (b) find relatively nonpolluting energy sources; (c) eliminate large dams except those built for water supplies; (d) stop introducing non-native, sport species to our watersheds; and (e) invest in management, control, and extirpation of other non-natives where feasible. Of course, making any of these happen is a very long shot and will take the efforts of many nature lovers working in concert!

Chapter 4
Leviathans and Lightning Wielders

Not All Giants Are in Fairy Tales

While the largest animal on this planet is the marine blue whale—which is a distant relative of terrestrial species like hippos—only a few small to large modern mammals and large reptiles live a significant portion of their daily lives in freshwaters. Included in this group are hippos, river dolphins, beavers, otters, muskrats, crocodilians, and giant snakes. Some of these build their domiciles in or very near freshwater systems (otters, beavers, and muskrats), while others spend a large amount of their day submerged or swimming in water (hippos). Another group rests near water and enter mostly to find prey (alligators, crocodiles, caimans, and many snakes). Some notes on many of these freshwater species are discussed in this chapter, but see also Chap. 5's coverage of beavers and muskrats and Chap. 1's discussion of otters.

In this chapter, I will also explore the unusual world of lightning wielders—those animals that can detect electrical charges—many of which can also deliver an electrical blow sufficient to stun or kill a nearby prey animal, potential enemy, or careless human.

River Horses

Only two species of hippopotamus—or "river horses" (from the ancient Greek name)—have survived into modern times. One is the species commonly seen in zoos and in the wilds of African rivers from the Congo to South Africa (*Hippopotamus amphibius*)—though previously their range extended to the Nile River—and tens of thousands of years ago, they were found in southern Europe. The related pygmy hippo (*Choeropsis liberiensis*; previously known as *Hexaprotodon liberiensis*) is

J. H. Thorp, *The Otter and the Fairy Shrimp*,
https://doi.org/10.1007/978-3-031-64029-2_4

Fig. 4.1 Herd of hippos (*Hippopotamus amphibius*) exiting a river. (Photograph courtesy of Flickr contributor Richard Probst)

largely nocturnal and is restricted to Liberia and small regions of other nearby countries in western Africa. Despite their name as river horses, hippos are not related to horses. Although it seems likely that whales evolved from ancestral hippos, paleontologists are currently stumped by the "ghost lineage" that led to the hippos themselves!

Members of the more widely distributed *H. amphibius* are gregarious but not truly social animals (Fig. 4.1). These hippos are typically found in small herds of 10–30 adults and offspring. During the day when they are not consuming large masses of aquatic plants, they are most commonly seen floating with only their eyes and nostrils above water, though occasionally they may rest on partially exposed sandbars. When completely submerged, they walk and more or less hop along the bottom rather than truly swimming. Adults can submerge for up to 5 min, but calves are limited to only about 30 s. Submersion provides better access to aquatic vegetation and also temporary protection from predators. At night, hippos often forage for food on land along well-worn paths but typically go only short distances from the water to lessen the chances of encountering dangerous predators. Despite their large body mass, they actually consume only about 88 lb/40 kg of plant matter per day on average, which equals about 1–1.5% of their body weight (still a large amount of plant material!). This may seem to be a large amount of food, but it is small compared to the typical daily consumption by a large terrestrial cow (2.5%). Their secret for not needing massive amounts of food is their largely sedentary lifestyle and living mostly in weight-supporting water.

The rarer and more threatened pygmy hippos display many of the aquatic adaptations of their larger cousins, but they spend relatively more time on land, where they typically mate and give birth. Their herbivorous diet is much more dependent on terrestrial vegetation (ferns, fruits, grasses, and various broad-leaved plants) than their larger cousins, but they still rely on periodic dips in the local water body to keep their skin moist and their body temperature cool.

Many people in North America and some other parts of the globe grew up with the story book or TV cartoon of the three little pigs (good guys!) and the wolf (bad guy!) and might therefore assume incorrectly that hippos are docile because of their outward shape which roughly resembles a big pig. However, they are not close evolutionarily, ecologically, or behaviorally—and actually, both domestic pigs and their wild boar cousins can be very dangerous. Indeed, despite the fact that river horses are vegetarians, they are one of the most dangerous animals in Africa because of their willingness to attack rather than flee, their strong tusks, and their huge size. Males can reach almost 10,000 lb/4536 kg, while females hit the scale at a not-so-svelte 3000 lb/1361 kg or less. During adolescence and within the mating season, territorial males in particular can be quite aggressive and may kill calves if they stray from their mother's protection. If her calf is threatened, a female tends to attack the more vulnerable side of the male rather than face his massive jaws head-on.

The greatest threat to these large and dangerous animals are humans who are seeking ivory for sale or are merely protecting their village homes and crops from hungry hippos. Water diversion via construction of dams and canals is a longer term and more permanent threat to hippos. Natural predators—which mostly attack the young—include crocodiles, various large cats, and packs of hyenas. While these take an annual toll on the population, this has generally not been a significant historical threat to the species as a whole but might become so in the future because of additional negative impacts of humans. In contrast, hippos have a useful albeit minor positive relationship with several species of fish which clean their hides.

In a strange and unfortunate twist occurring in the 1980s, a drug lord introduced two hippos into the Magdalena River in the South American country of Colombia. Estimates of the current population are around 130, with the numbers continuing to grow at a rate of perhaps 10% per year! In Columbian rivers and along the shore, they are destroying plants, polluting the waters with their abundant wastes, and threatening wildlife and even people. As a rather questionable policy, the Columbian government has started shipping live hippos to Mexico and India! It is hard to imagine a more environmentally damaging and dangerous species to introduce to a river system!

Those Cold-Blooded, Reptile Predators

Crocodiles and Alligators   While hippos tip the scale (or more likely crush it) as the largest freshwater species, right behind them in size and danger are the semiaquatic crocodiles and to a lesser extent alligators and caimans, which surprisingly enough

Fig. 4.2 Nile crocodile (*Crocodylus niloticus*). (Photograph courtesy of Flickr contributor Martinus Scriblerus)

are the nearest living relative to birds! Crocodiles, alligators, and caimans are carnivorous members of the taxonomic order Crocodilia. The largest and most dangerous to humans is the very aggressive Nile crocodile (*Crocodylus niloticus*; Fig. 4.2), which despite its name is actually widespread in sub-Saharan Africa where it is responsible for hundreds of human deaths each year. The deaths have come from people traveling on waters and land, because the crocodile can run very fast (20 mph/32 kph) for short distances—which is faster than the average human can run. The problem on land, however, is not being run down in a long race but instead being surprised by a croc that launches itself a short distance from ambush. Individual reptiles have been killed that reached over 20 ft/6.1 m in length and weighed over 2400 lb/1089 kg, making this freshwater crocodile second in size only to the saltwater crocodile among reptiles. Members of the order Crocodilia can be found in rivers, lakes, and wetlands. [Several saltwater crocodiles have been captured in South Florida in recent years, which is a very frightening prospect for people visiting the Florida Everglades; so far, these seem to have come from the action of one very irresponsible human.]

While crocodiles have a bad reputation from the human perspective, they are actually aiding humans and nature in the Northern Territories of Australia by substantially reducing the population size of introduced and now feral pigs that are damaging the wild terrain and helping accelerate Australia's very high rate of terrestrial mammal extinctions. After crocodile hunting was banned there in 1971, the number of crocs expanded from a near extinction level of 3000 to the current estimated level of 100,000. One developing response has been the desired reduction in feral pigs.

Alligators are closely related to crocodiles but are smaller and less aggressive— nonetheless, attacks on and injury or deaths of humans occur yearly even in populated areas of the southern USA. While these are no doubt very dangerous animals, on average only about one human fatality per year occurs in the USA from alligator attacks. In comparison, bees, hornets, and wasps cause just over 50 deaths per year in the USA. "Gators" are also hard on dogs that go swimming in a lake with these reptiles. In marked contrast, it is estimated that 20–25,000 alligators are harvested annually just in the single US state of Louisiana. Alligators are sought for their meat and hides, the latter of which are used for boots and other items of apparel. The American alligator (*Alligator mississippiensis*) is fairly common and extends northward to North Carolina (albeit very rare) in the USA but is especially common in the Deep South. A critically endangered but less aggressive species is the Yangtze alligator of China (*A. sinensis*). Oddly enough, this reptile may have been the inspiration for ancient stories of dragons in China.

Two other taxa in the order Crocodilia that occur in freshwaters are the caiman and the gharial. The former occurs in Mexico and both Central and South America. Most are relatively small (13–88 lb/5.9–40 kg) and seem less aggressive than their larger relatives, but the black caiman (*Melanosuchus niger*) of the Amazon can reach 13 ft/4 m and weigh 1100 lb/499 kg. Gharials (*Gavialis gangeticus*) of the Indian subcontinent are noted for their very long and narrow head, and their total body length for males is in the range of 10–20 ft/3–6 m. They tend to spend a greater portion of their time in the water compared to other crocodilian species.

Stealthy Serpents Aquatic snakes are also common reptiles found in and around the world's freshwaters. They can be found throughout the world in temperate to tropical latitudes, but the largest are in the tropics. Some kill their prey with venom, while others merely crush and then swallow them whole without using venom. Perhaps best known to the television-watching public are the anacondas, with the largest of these being the green anaconda (*Eunectes murinus*) (Fig. 4.3). These massive snakes can reach nearly 30 ft/9.1 m in length, 550 lb/249 kg in weight, and over 12 in/30.5 cm in diameter, making them the world's largest snakes.

Anacondas are found primarily in swamps and marshes and slow moving streams within rainforests of the Amazon and Orinoco basins of South America. They lay quietly in streams with only their eyes and nostrils above water, waiting for a wild pig, a deer, or even a jaguar to venture near to them. They then crush or merely suffocate their prey with their strong coils and employ stretchy ligaments in their head that allow then to swallow their prey whole. Such large meals combined with the snake's generally quiescent behavior permits an anaconda to go for weeks or even months without a second meal.

At least ten species of semiaquatic snakes frequent North American waters, especially in the southern USA, with the vast majority posing no threat to humans. The most dangerous, however, are the venomous water moccasins, also called cottonmouths (*Agkistrodon piscivorous*) (Fig. 4.4). While these usually hunt in the water,

Fig. 4.3 Green anaconda (*Eunectes murinus*). (Photograph courtesy of Flickr contributor Ferando Flores)

Fig. 4.4 Cottonmouth snake (*Agkistrodon piscivorous*). (Photograph courtesy of Flickr contributor at the Virginia State Parks)

they often rest on a log, stump, or bankside vegetation until prey passes near. Although these are far smaller than anacondas, they typically reach at least 2–4 ft (61–122 cm) in length. While working at the Savannah River Ecology Lab near Aiken, South Carolina, I frequently encountered cottonmouths from boats while doing research in cypress swamps of the southern USA. I once faced one (literally) curled up on the bank as I was preparing to climb out of a roadside canal from a trip to collect crayfish from baited traps—an encounter that nearly scared the bejesus out of me!

Kindly Turtles … More or Less Compared to snakes and many other animals, turtles have the proper reputation of being somewhat innocuous in most cases, but that does not mean you should be entirely careless around them. The vast majority of readers of this chapter will have only encountered live turtles in a pet store or may have seen or collected small ones from a local pond, wetland, or stream. [We are properly ignoring here the giant sea turtles—as interesting as they are—and will focus instead on the denizens of the freshwater world.] Most of what the average reader will have seen are hard shell turtles, but there are many softshell turtles in rivers around the world, some of which are quite large.

Softshell turtles (family Trionychidae) do not conform to the common vision of a turtle that many people have, and some of these reptiles can be quite large. Their more leathery shells, or "carapace," usually lack the hard horny scutes (or scales) we associate with most turtles, but the central area of the carapace still contains a bony support. Consequently, these shells are lighter and more flexible than those of other turtles, which enables the turtle to move faster on land and in the water. Females are typically larger than males by 25–50% in shell length. As poikilotherms (= animals mostly unable to regulate internal body temperatures), both sexes spend much of their day lying in the sun but will retreat to water if threatened and may lie there buried mostly in the sand until unduly disturbed. I once was sitting near shore in the shallow water of the Kansas River looking at an aquatic insect I had collected when suddenly a softshell turtle "sprang" from the sandy waters only 2 feet away from me as it apparently chose not to risk staying any longer! I am not sure which of us was more surprised!

One of the most unique features of softshell turtles—in addition to their leathery shells—is that they can "breathe" underwater while lying motionless because their mouth contains tissue filaments richly supplied with blood vessels to absorb oxygen from the environment. This system would not work adequately for active turtles. Staying underwater for longer periods enables these turtles to capture more prey from a menu of aquatic insects, crayfish, and occasionally small fish.

A giant softshell turtle from Southeast Asia (*Pelochelys cantorii*) (Fig. 4.5) is critically endangered through most of its range, which includes the Mekong River. Its carapace is generally 28–39 in/71–99 cm long, and this behemoth often tips the scales at 220 lb/100 kg. Like most softshell turtles, it lies buried for the greater part of the day in the sand of streams and rivers waiting to capture small fish, snails, crabs, and shrimp that come within range of its protrusible neck and jaws. The Yangtze giant softshell turtle (*Rafetus swinhoei*) of China and Vietnam is

Fig. 4.5 Giant softshell turtle (*Pelochelys cantorii*). (Photograph courtesy of Flickr contributor Dementia)

comparable in size, but it is in an even greater danger of extinction. The greatest threat to softshell turtles around the world (including in the USA) comes from habitat loss and consumption of turtle flesh by humans.

Hardshell turtles can also reach behemoth sizes and tend to be much more dangerous than their softshell cousins. The alligator snapping turtle (*Macrochelys temminckii*) is the largest turtle in North America. It occurs in rivers and swamps of mostly the southeastern USA, though its current, reduced range does extend northward into several midwestern states. It has prominent ridges on its shell (Fig. 4.6) and powerful jaws that can snip off fingers from incautious humans attempting to capture it. Most adults reach at least 14–32 in (36–81 cm) in shell length and generally can reach 176 lb/80 kg in weight. Unlike the softshell turtles discussed above, males of these hardshell turtles are bigger than females. They eat various fish and aquatic invertebrates but will also consume some small to medium-sized terrestrial

Fig. 4.6 Alligator snapping turtle (*Macrochelys temminckii*). (Photograph courtesy of Flickr contributor Sam Stukel at the US Fish and Wildlife Service)

vertebrates unlucky enough to attempt to swim across the stream where these turtles wait in ambush. The Arau turtle (*Podocnemis expansa*) of the Amazon Basin in South America grows slightly larger, reaching up to 200 lb/91 kg. However, its diet is very different from the alligator snapping turtle in that it feeds primarily on aquatic vegetation.

Giant Salamanders and Frogs

To most of us, salamanders—a name derived from the Greek term meaning "fire lizards"—are small vertebrates that we might find in damp areas along or in streams and wetlands and even in caves (where many are blind; see Chap. 9), so it may seem odd to discuss them in a chapter on behemoths. Indeed most species are only 6 in/15 cm long, while the pygmy salamander may reach the minuscule length of only 0.6 in/1.5 cm. In contrast, giant salamanders in the genus *Andrias* (family Cryptobranchidae; Fig. 4.7) can attain a length of nearly 6 ft/1.8 m! They are native to Japan and southeastern China (assuming that the latter species still survives in nature) and have been known to live for 50 years in captivity. In the wild, these amphibians consume aquatic insects, frogs, crabs, shrimp, and even small fish. A male builds and occupies a small den in a permanent stream, usually welcoming only sexually active females to join him. Local legends have described this salamander as having magical regenerative powers, while other rumors have it eating small children and livestock! In contrast, the largest amphibian in North America is the hellbender salamander (*Cryptobranchus alleganiensis*) found in the eastern USA (Fig. 4.8), but its maximum length is only 29 in/74 cm.

Fig. 4.7 Giant salamander (*Andrias japonicus*). (Photograph courtesy of Flickr contributor Marshal Hedin)

Fig. 4.8 Hellbender salamander (*Cryptobranchus allegani*). (Photograph courtesy of Flickr contributor Brian Gratwicke)

By the way, confusion often arises about the difference between a newt and a salamander, in part because all newts are salamanders but not all salamanders are newts! Of the two, newts spend most of their lives in freshwaters, while salamanders mostly appear there primarily when breeding and producing eggs. While both require at least humid habitats, the paddlelike tail and webbed feet make the newt more adapted to life in a small stream or pool. The skin of a newt is toxic to most predators, thereby providing them some safety from the many predators inhabiting forests.

Aside from salamanders, some very hefty giant frogs can be found on temperate and tropical continents of the world, but none reach more than about 12.6 in/32 cm long and 7.3 lb/3.3 kg in weight. The largest is the Goliath frog (*Conraua goliath*) on the west coast of equatorial Africa. It is so large that it can move large stones weighing half its weight to create a dammed pool by sandy riverbanks where its tadpoles can be hatched and grow for a time in this aquatic nest.

Colossal Fishes

Sturgeons and Paddlefishes   Usually the largest fish the average person has seen— and no doubt recounted in numerous "big fish" stories—is the one that you almost got into your boat before it broke the line! But aside from the really big stingrays to be discussed in Chap. 6, many other giant freshwater fish exist in the temperate and tropical zones of the world that are worthy of discussion without the need to cross your fingers behind your back when you describe them!

"Sturgeon" is the common name for a fish family with up to 27 species that evolved with relatively little changes over the last 200 million years. Some are entirely residents of freshwaters, but others are anadromous and thus migrate from estuaries or from the ocean to breed in freshwater. Nine species are endemic to North America, all of which are classified as endangered, threatened, or species of concern. The most common in North America is the lake sturgeon (*Acipenser fulvescens*); it spends its entire life in freshwaters. These primitive species have changed very little over this period and are noted anatomically for their mostly cartilaginous skeleton, which contrasts with the bony framework of most freshwater fishes. They also differ from most freshwater fish in having a tail (caudal) fin (Fig. 4.9) which resembles that possessed by many sharks. They occur within the northern hemisphere in rivers and lakes from the subarctic to the subtropical latitudes of Eurasia and along both coasts and inland large rivers of North America. Most species are bottom-feeders on mussels, crustaceans, and small benthic fish. Sturgeons use electroreceptors in their head to detect weak electric fields produced by potential mates as well as by prey animals. They can also identify geoelectric sources on the river bottom that may contribute to migration efficiency. The largest species can reach 18 ft/46 m long and weigh 4400 lb/1996 kg. Scientists have determined that some individual fish can attain an age of 100 years, with most not becoming sexually mature until they are 20 years old and most living only 50–60 years on average

Fig. 4.9 Sturgeon (*Acipenser* sp.). (Photograph courtesy of Flickr contributor ramendan)

unless killed earlier by humans. This slow growth and delayed maturation when combined with a very unique value (see below) make them highly prized by humans and easily subject to extinction. As a result, over 85% of sturgeon species are now close to extinction, making them one of the most endangered groups of freshwater fish species worldwide. This depletion in population size poses significant cultural and dietary problems for some Native American nations, such as the Menominee people of Wisconsin who have relied from ancient times on these fish as important components of their diet.

What is that special value to humans that threatens this ancient fish? The answer is simple and can be directly attributed in great part to the delicacy known as "caviar"—which is just a fancy and more savory name for "fish eggs." Caviar, which can be an acquired taste, consists of salt-cured roe (eggs) of sturgeons (and to a lesser extent roe from paddlefish and a few other fish species). The rarest and thus most expensive is beluga caviar, which is noted for the large size, softness, and savory taste of the roe. For example, connoisseurs—or just plain very rich folks—were paying at least $200–300 US for an ounce of beluga caviar in restaurants in the fall of 2022.

Related to sturgeons, but very different ecologically, is the equally ancient paddlefish of North America (*Polyodon spathula*; also called spoonbills; Fig. 4.10) and the Chinese paddlefish or swordfish (*Psephurus gladius*), which may now be extinct. Adults of the North American species average almost 5 ft/12.7 m in length and weigh up to about 60 lb/27 kg. This species is a zooplanktivore in both larval and adult stages, making it the largest fish in North America that depends almost

Fig. 4.10 North American paddlefish (*Polyodon spathula*). (Photograph courtesy of Flickr contributor Ryan Hagerty at the USFWS)

exclusively as adults on zooplankton (cladocera, copepods, and probably fish larvae) as a food source. Unfortunately, these same zooplankton depend on planktonic algae, which are also heavily consumed by invasive zebra mussels, thereby reducing the abundance of zooplankton for the paddlefish. Although the most expensive caviar in the world comes from sturgeon, the eggs of paddlefish are also sold as caviar, leading to the decline in paddlefish abundance.

Lurking beneath the waters of North America is another primitive fish, the alligator gar (*Atractosteus spatula*) (Fig. 4.11). The largest recorded this century in North America was nearly 8.5 ft long (2.6 m) and weighed about 330 lb/150 kg. They have a double row of sharp teeth that are employed to impale and hold their prey, which is usually another fish or crayfish but could also include waterfowl and small mammals swimming on the surface. Gars can temporarily survive in low oxygen conditions by sticking their heads out of water and gulping atmospheric oxygen. Some fishermen will unfortunately kill these fish and toss them away, with the erroneous view that they are bad for the aquatic community and less edible. When inexpertly handling gars, a person may run afoul of the unusual, bone-like, ganoid body scales with their serrated edges that can accidentally cut an angler if the fish is rubbed from the rear forward.

Giant Tropical Fishes Aside from the freshwater sharks and stingrays discussed in Chap. 6 and the sturgeons, gars, and paddlefish described above, many other giant fish lurk beneath the surface of the world's waters.

Fig. 4.11 Giant alligator gar (*Atractosteus spatula*) collected in 1910 from the Mississippi River. (Photograph posted by courtesy of Flickr contributor David Foster)

The migratory Mekong giant catfish (*Pangasianodon gigas*) is in a taxonomic group of fishes called "shark catfishes" that differ in shape and evolution from other large catfishes in the north temperate zone. Reaching a length of nearly 10 ft/3 m and 330–440 lb (150–200 kg) in just 6 years, it is the heaviest and fastest growing, entirely freshwater fish species, or at least it is a strong competitor in that weight class to some sturgeons. Despite its imposing and potentially threatening size, it feeds mostly on detritus and algae it finds along the river bottom. The distribution of this giant endangered fish is now limited to medium to large rivers within the middle portion of the Mekong River.

Another of the large aquatic animals is the *Arapaima* (also called either by the Portuguese-Tupi name "pararcu" meaning "red fish" because of their tail color or the Peruvian name paiche) (Fig. 4.12). These tropical fish of South America can be seen in many large, public aquaria around the world. Like the catfish described above, the arapaima can also attain a length of about 10 ft/3 m, but they are much more streamlined than the Mekong giant catfish. The arapaima has the unusual ability to breathe both in and above the water, with the latter technique dependent on a modified swim bladder that enables them to survive up to 24 h in the deoxygenated pools and streams which they commonly find themselves. They typically feed on fish but may also eat terrestrial insects and small vertebrates along with fruits and seeds that have fallen into the water. Their lifespan of about 20 years is shorter than many other really large fish.

The common river and lake fishes known in the USA under the general name "carp" are members of the taxonomically huge family Cyprinidae. That taxa contain many native species such as the highly diverse and abundant, true minnow species. The fish North Americans call carp were first introduced from Europe hundreds of

Fig. 4.12 Giant *Arapaima gigas* (also called pirarucu or paiche) from the Amazon Basin. (Photograph courtesy of Flickr contributor "Haka's photos")

years ago. Today individual carp, some which have been given personal names by local sports anglers in Great Britain, are typically caught and then released to be captured again at a later date. Once when I was working in some side channel pools of the St. Lawrence River in upstate New York, I observed two Englishmen fishing for carp—something almost no Americans would consider doing, as they are often called "trash fish." One thing that impressed me was the very expensive and elaborate fishing gear that they were using—clearly they were not seeking carp for an evening's meal!

In the last few decades, our long-existing though non-native North American carp were intentionally supplemented with several new and very invasive species from China, including the black, bighead, grass, and silver carp species. In most cases, they were supposedly confined to aquaculture ponds—a process that later proved disastrous to our resident species when they escaped! These new emigrants have generally had a very negative effect on our native species and have proved almost impossible to control, even by electrical barrier erected by the US Army Corps of Engineers. The densities of these carp can be huge in our rivers and lakes. When large groups are disturbed by a passing motor boat, many carp jump up into the air. This has resulted in minor to very serious injuries to people in the boat when the almost inevitable, human-carp physical contact ensues.

While these current US carp fish generally attain a large size, the most massive species of carp worldwide is the endangered "giant barb" or "Siamese giant carp" (*Catlocarpio siamensis*) from Southeast Asia which lives in swamps and shallow streams. They originally were known to grow to a length of ~10 ft/3 m, but smaller species are now collected. Current individuals tipping the scales at 660 lb/299 kg can weigh much more than either the Mekong catfish or the arapaima. The giant barb eats benthic algae, phytoplankton, and fruit. While their numbers in nature have been decimated, this species is now being raised in aquacultural ponds for commercial sales and restocking into local rivers.

Gargantuan Crustaceans

As a brief closing to this coverage of freshwater leviathans, I would be remiss without mentioning the three species of giant freshwater crayfishes or lobsters found on the island of Tasmania, including *Astacopsis gouldi*. This unusually large freshwater crustacean—which looks more like a crayfish than a lobster—is known as "lutaralipina" in the aboriginal language. Not only is it the largest crayfish on Earth, but it is also the heaviest freshwater invertebrate in general on our globe, with most adults reaching 4 to almost 7 lb (1.8–3.2 kg) and stretching to just over 30 in/76 cm! At that huge size, this crayfish has no natural enemies (other than humans)—except at a very young age when they may be consumed by a passing platypus or a very large fish. When undisturbed, this omnivorous crustacean can live for up to 60 years in Tasmanian rivers where it can be found at elevations of 1300 ft/396 m, or more. It is legally protected in Tasmania and is on the IUCN List of Threatened and Endangered Species. [To remind you, IUCN is the abbreviation for the highly important "International Union for Conservation of Nature"—a worthy organization to support.]

Two other crayfish from this region of the planet are also somewhat massive (Fig. 4.13a, b), though not as large as the Tasmanian behemoth. The Murray River Crayfish (*Euastacus armatus*) of southern Australia (Murray-Darling Basin) can attain a weight of 5.5 lb/2.5 kg, while the marron (*Cherax tenuimanus* and *C. cainii*) may reach 4.9 lb/2.2 kg. The marron and two other *Cherax* species (the red claw crayfish and the yabby) have been the focus of aquaculture production as a food source for Aussies because of the crayfish's large size and succulent nature. Some

Fig. 4.13 (**a**) Large Australian red claw crayfish (*Cherax quadricarinatus*) (Photograph courtesy of Flickr contributor "Nature.Catcher"); (**b**) large Murray River crayfish. (Photograph courtesy of James H. Thorp)

attempts have been made to develop aquaculture operations for these crustaceans on other continents, but this is an especially bad idea from an environmental perspective because escaped individuals could possibly devastate our native and smaller North American crayfish and also prey on our native community of freshwater mussels. In contrast, the slower growth and more aggressive behavior of the Tasmanian giant crayfish make it unsuitable for aquaculture.

For more information on crayfishes of the world, see Chaps. 2 and 5.

A Stroke of Bad Luck?

Many of us were told by our parents not to stand under a tree in a lightning storm, but did they warn you about fish that can also deliver a shocking blow? Probably not if you live in the northern hemisphere, but there are electric fish (electrogenic fish) in the ocean (such as torpedo rays) and in freshwaters of Africa and South America. Many fish can detect electric fields, but only a few can produce enough electricity to stun a potential prey or predator. Most produce only weak electrical impulses, but some generate a very strong charge sufficient to stun an adult human. The shock can be equivalent to or several times greater than the shock generated by accidental contact with an electrical outlet in your home. If that person has a weak heart or is in deep enough water to drown, death may result from near contact with these fish. In general, "bioelectrogenesis" is employed for purposes of hunting prey, self-defense, electro-communication, and navigation in murky waters (much as terrestrial bats do at night by other means). This capacity has been evolved independently in six lineages of fish. Among freshwater fishes able to generate an electric field are ghost knifefishes, electric eels (closely related to knifefishes), and electric catfish. The last two groups can generate very strong electric charges.

The most powerful electrogenic fish is found in freshwaters of the northern portion of South America. *Electrophorus voltai* is called the Volta's electric eel or arima, meaning "something that deprives you of motion." And to make matters even more dangerous, this Volta's electric eel hunts in packs as large as 30 eels in their pursuit of fish prey! An individual fish can be around 8 ft/2.4 m long and weigh 44 lb/20 kg. It can also generate up to 850 V of electricity—which is about 250 V more than its nearest contender! Their bodies are slender, and their heads are somewhat flattened. Another electric eel (actually not a true eel but in a related group of knifefishes akin to catfishes) is the better known and more widely distributed *Electrophorus electricus*. It occurs in the Amazon and Orinoco River basins of northern South America where it lives in areas with muddy bottoms and dense vegetation. While it feeds primarily on invertebrates, it may occasionally consume fish and rats. Waters of central and western Africa are home to the electric catfish (*Malapterurus electricus*), which can grow up to ~3 ft/0.9 m long (Fig. 4.14).

Fig. 4.14 Electric catfish (*Malapterurus electricus*). (Photograph from the Steinhart Aquarium in San Francisco and courtesy of Stan Shebs)

How Do They Do It?

The former "magical" secret to how electricity is produced by these fishes has now been exposed. A large network of modified muscle cells formed into stacked, prism-like structures and associated connective tissue (called electroplaques) generate the electrical charge through a simultaneous depolarization in series and parallel of the electroplaques. This in-series alignment of the muscle cells and thousands of electrolytic muscle cells produce an additive voltage similar to the effect of stacking batteries in a flashlight.

Other Electric Users: Australian Platypuses

One of the strangest mammals on this planet is the semiaquatic platypus *Ornithorhynchus anatinus* (Fig. 4.15). It is evolutionarily in a very small phylogenetic group with other monotreme species consisting of the four distantly related species of echidna, or "spiny anteaters." Its oddity is that females possess mammary glands while also laying eggs in underground burrows, and both genders have venom for protection! So unique is this species that scientists when first viewed a preserved specimen in England in the late 1700s felt that this was an elaborate hoax. The platypus has been assigned a near-threatened status by the IUCN in its range within eastern Australia and Tasmania. Despite its small size of 1 lb 9 oz to 5 lb 5 oz (0.7–2.4 kg) , the males of this species when threatened can pack a major wallop

Fig. 4.15 Semiaquatic platypus (*Ornithorhynchus anatinus*) from Tasmania. (Photograph courtesy of Flickr contributor Klaus)

with the venom injected from their rear ankle spurs. A human foolish enough to carelessly handle one of these aquatic/terrestrial mammals may experience extreme, temporarily paralyzing but nonlethal pain.

Aside from those well-known characteristics of platypuses, the reason they are described in this chapter is that their bill contains 40,000 electrical sensors arranged in a series of stripes. This allows them to passively locate living prey on the bottom of streams and ponds and distinguish them from inanimate objects. Their abilities to locate prey electrically is superior to that found in electric fishes.

A Historical Note on Animals and Electricity

If you find the subject of animal-generated electricity of interest, you will find yourself in a long line of previous humans who have been fascinated with this subject. In a shocking aspect of early medicine, the early Egyptians employed marine electric torpedo rays as an electro-therapeutic treatment for epilepsy, and the ancient Greeks also employed them in medicine. Likewise, the well-known statesman, author, and politician Benjamin Franklin, who was famous for investigating lightning with kites, also studied the generation of electricity in marine torpedo rays. But for real fun, you would have needed to attend some Victorian parties where excitement was generated among the guests by linking hands in a chain with one person at the front touching an electric ray with shocking results for everyone in the chain—man, did they know how to have fun in those days!

Chapter 5
Busy Beavers and Other Aquatic Architects

The architectural structures built by animals are often a favorite subject of nature enthusiasts and numerous book authors, but almost the entire focus has been on terrestrial animals from termites to birds. The only aquatic-related animal to receive comparable attention is the beaver. In this chapter, I will take a different approach and start with small aquatic insects and their creations (specifically the home-cases and capture nets of caddisflies), shift to a slightly larger but less complex invertebrate architectural features (crayfish burrows), switch to small vertebrates that build modest-sized structures (muskrat burrows and lodges), and end with the largest of all freshwater edifices—the lodges and dams built by beavers.

Caddisfly Homes: Cases and Nets

Net-spinning caddisflies are members of the ancient and very successful aquatic insect family Hydropsychidae in the order Trichoptera, a name meaning "hairy wings" (in reference to the long silky hairs on adults). They occur in almost all unpolluted, surface freshwater habitats on six continents as one of a limited group of insects along with a diverse group of arachnids that produce silk for various purposes. Caddisflies are closely related to the sister group of insects that is most closely associated over the centuries by humans with producing silk—the butterflies and silkworm moths in the order Lepidoptera. Caddisflies are also known to many fly-fishing anglers as "sedge flies" or "rail-flies."

Silk is employed for two primary purposes by caddisflies: (1) construction of sturdy, often motile cases for the larva initially and later for the metamorphosing pupa, with the cases typically characteristic of a specific taxonomic family within the order or sometimes just a genus within the family, and (2) weaving fixed capture nets and retreats for species that filter the water for food. The most common feeding strategy is to be a collector-gatherer where the caddis crawls around the stream

J. H. Thorp, *The Otter and the Fairy Shrimp*,
https://doi.org/10.1007/978-3-031-64029-2_5

Fig. 5.1 Protective case-like home of one species of larval caddisfly (*Pycnopsyche gentilis*). These cases vary considerably in design among taxonomic groups. (Photograph courtesy of Flickr contributor Bob Henricks)

bottom while protruding slightly from its protective case in search of food. The nutritive goals consist of obtaining and consuming small animals, algae, and sometimes dead organic matter (detritus). Other caddisflies are shredders that tear up organic matter into edible-sized pieces or scrapers that remove algae from rocks and wood surfaces.

The architecture and building material in a caddisfly's case can often reveal the family and sometimes the species that built the case (Fig. 5.1). Differences among architectural plans include the type and diversity of materials used to construct the cases, which depending on the caddisfly species may include minute pebbles, sand grains, wooden sticks, and even tiny but empty snail shells all woven or "glued" together with silk. In addition to variation among taxa in construction material, prominent differences may occur in how the items are arranged on the outer wall of the case. The interior of the case is formed entirely from silk. Some artists have taken advantage of this tendency by limiting the caddisfly in the lab to preselected material for their mobile homes, including providing the insect with only a choice of colorful stones, pearls, and even small amounts of gold and then photographing the resulting product or in some instances selling modifications of the eventual case as jewelry (Fig. 5.2).

Other caddisflies adopt a different strategy for finding food by building nets associated with an attached retreat (Fig. 5.3) in order to capture prey drifting downstream. The mesh size reflects the size of the desired food item, the caddis species

Fig. 5.2 Caddis cases in which artists have given the caddisflies only pieces of gold and pearls to construct their homes. (Photograph courtesy of Hubert Duprat in Venice)

Fig. 5.3 Caddisfly nets in a stream. (Photograph courtesy of Flickr contributor Mike Majeski at Emmons Olivier, Inc.)

building the net, and sometimes the surrounding water velocity and turbulence. If the species eats drifting algae, then a small mesh size is required. However, if the desired prey is a small animal, then a larger mesh is more appropriate as it allows more water to pass through the net, thus increasing chances of catching a drifting insect prey. A larva will regularly clean its net to maintain ideal flow rates while also extracting and consuming any desirable prey items. It will also capture any suitable prey possible that have bypassed the net but have become enmeshed on its nearby, downstream home where the caddisfly attempts to remain safely hidden from predators for most of the day. As the caddisfly approaches pupation and eventual metamorphosis into a terrestrial adult, it becomes vital to capture some animal prey with their higher energetic value. Consequently, the late instar caddis will often clean other material from its net that it may have eaten when in an earlier instar stage so that water flow and potential capture rates increase.

One hazard of moving about the stream or lake bottom is that the caddis is more exposed to predators. Common defensive tactics are to be nocturnal and/or to remain hidden in masses of decaying leaves and other organic material. Another supplementary approach used often by constructors of immobile retreats is to produce warning stridulations in an attempt to ward off small predators or with the hope of scaring off a competitor seeking to steal their case. These vibrations are produced by rubbing bumps on the caddisfly's anterior femurs. How effective this action is over time has not been well established.

Burrowing Crayfish

Most people in both rural and urban environments are at least vaguely familiar with crayfishes (Crustacea, Decapoda), which also may be known locally in the USA as crawdads, crawfish, craydids, or mudbugs. They are more distantly related to crustaceans in other continents called freshwater lobsters, mountain lobsters, rock lobsters, baybugs, or yabbies. Where individual and multispecies assemblages of crayfishes are typically found reflects (a) regional climate, (b) abundance and type of available aquatic environments, (c) accessibility to freshwater rivers and lakes, (d) presence of competitors and predators, and (e) choice of available aquatic habitats (lotic, lentic, and burrowing in moist soil). While most people probably think of crayfishes as an entirely aquatic animal—if not as a menu item in a Cajun restaurant—crayfish experts (or "astacologists") would tell you that some species are "ecosystem engineers" that excavate burrows on land. These burrows extend from the soil-air interface downward to a shallow water table, where the resident crayfish employs the underground water for general body moisture and to keep its gills wet for breathing. It is also a place to invite a partner for a little "socializing" with the goal of producing more crayfish! The tendency to burrow and the skill at doing so increase in frequency from rarely burrowing crayfish (= tertiary burrowers), to occasionally burrowing species (= secondary burrowers), to species that are always found in burrows (= primary burrowers), as do the extent of anatomical adaptations.

All choices of where an animal resides comes with costs and benefits. In the case of crayfish, living in a totally aquatic habitat like a stream or lake provides greater accessibility to food and mates, but it comes at the expense of greater exposure to aquatic and terrestrial predators and sometimes dangerous water currents during floods. An alternative choice for some species is to build a burrow on moist ground close to a water body where safety is emphasized at the cost of a lower access to food sources. Such burrows may penetrate downward somewhat over 12 in/30.5 cm to as much as 15 ft/4.6 m. Medium to large burrows can be relatively complex architecturally and involve multiple chambers and tunnels, underground exits (including to the adjacent stream in rare cases), and short-to-tall chimneys in most cases (Fig. 5.4). The purpose of the soil-based chimney system is to help passively regulate airflow by drawing air inward at a lower-chimney entrance and having it pass outward through a second and taller chimney. This design is well-proven in the animal kingdom from terrestrial prairie dog tunnels to some tube-dwelling marine polychaete worms (e.g., *Chaetopterus*).

Species of primary burrowing crayfish stay in their burrows almost all the time except to find food and a mate and then sometimes only venturing forth during the

Fig. 5.4 Crayfish chimney in the North Carolina wetlands. (Photograph courtesy of Flickr contributor NC Wetlands/ Kristie Gianopulos Yates)

rainy season and at night when predators may be fewer. When searching for food, the crayfish may travel to a local water body or just wait at the entrance of their tunnel to catch some suitable prey animal passing near the burrow opening. In cases of prolonged drought, burrowing crayfish often seal their burrows to retain its vital moisture. Because these crustaceans rarely have to resist water currents in a local stream, their abdomen and its flexure muscles tend to be slimmer than those of their stream inhabiting cousins. In contrast, they tend to have relatively larger gills and bigger thoracic chambers to house those gills.

Crayfish burrows may provide habitat for other animals, especially during droughts. These can include many species of insects, spiders, nematode worms, amphibians, snakes, and small mammals. For example, Hine's emerald dragonfly seeks shelter there in winters and in very dry periods in other seasons. If you are considering seeking a crayfish from a burrow, be cautious—I have helped excavate burrows in South Carolina in search of crayfish and can guarantee that a concern about finding a rare water moccasin hiding in the burrow was always on my mind!

Muskrat Lodges

Muskrats (*Ondatra zibethicus*) are medium-sized, semiaquatic rodents native to North American wetlands (Fig. 5.5) that have been introduced to at least three other continents (Asia, Europe, and South America). They are relatively small (1.3–4 lb/0.6–1.8 kg) compared to beavers, with an average length of 8–10 in (20–25 cm). Despite their name, muskrats are not closely related to rats, although they are in the rodent taxonomic order Rodentia (the largest mammalian order). Their family is Cricetidae, which also contains hamsters, voles, and some mice. While they are in a different taxonomic family from beavers (Rodentia, Castoridae), they share some ecological similarities because of their association with water

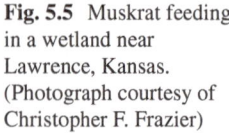

Fig. 5.5 Muskrat feeding
in a wetland near
Lawrence, Kansas.
(Photograph courtesy of
Christopher F. Frazier)

bodies and their tendency to construct lodges, albeit not as complex and sturdy as those built by beavers. Although they have long, hairy, and slightly flattened tails, these tails are again not modified to the extent shown by beavers with their broad, scaley, and greatly flattened tails. A swimming beaver often only shows its head above water, while a muskrat's head, back, and tail can usually be seen on the water's surface.

The common name of muskrats comes from their odor, which has been described as a musky smell. Beware! Knowing the glandular location of this smell and the proper technique to remove this organ are critical steps in preparing muskrat stew for your dinner meal!

While only beavers make dams, both muskrats and beavers construct lodges when not living in burrows within the bank (see description of beaver lodges in the next section). It is easy to discern which of these two rodents are present because of their distinctive lodges visible from the bank. The normal muskrat lodge (Fig. 5.6) is smaller and is mostly constructed with cattails and other reeds rather than with sticks and logs, and thus does not have the longer lifespan of an undisturbed beaver lodge. Lodges of both muskrats and beavers can end up clogging waterways, leading to humans removing them, with an ultimate negative and sometimes lethal cost to these fascinating mammals. In contrast, muskrats are helpful to humans because they eat aquatic plants (as well crops sometimes), thereby providing open spaces for waterfowl to land and swim. Their lodges also provide on their upper surface nesting sites for waterfowl and resting places for transient birds.

Muskrats can be easily confused with another aquatic mammal, the "nutria" (*Myocastor coypus*), which are also called coypu or beaver rats and by other names on different continents. These are large rodents originally from South America. Nutria were intentionally introduced to parts of the world for potential fur production but later escaped or were intentionally released. In North America, these semi-aquatic mammals now reside in many coastal areas including the northwestern and

Fig. 5.6 Muskrat and its lodge. (Photograph courtesy of Flickr contributor "Putneypics")

Gulf Coastal states. Nutria (Fig. 5.7) are intermediate in size between muskrats and beavers, but they do not build lodges.

Beavers: The Master Architects and Builders

Distribution and Anatomical Portrait of Beavers

I am ending this chapter with coverage of the master builders of the freshwater world—the beaver, including *Castor canadensis* (Fig. 5.8) as found in Canada and the USA, and *Castor fiber*, a smaller-bodied species native to Eurasia. They are notable as being the largest rodents on our planet, except for two species of capybaras (*Hydrochoerus*) from South America, Panama, and the nearby island of Grenada—with the larger of these two tropical species reaching ~143 lb/65 kg. Unlike the beaver, however, capybaras are not accomplished builders. The North American beaver by contrast is less than half that size, reaching only 24–66 lb (11–30 kg). Beavers are widespread in Canada and the USA in areas where sufficient trees grow to provide the vital materials for building their dams and lodges. So common is this animal throughout much of the wild and even rural areas of North America that it has been designated as the "national animal" of Canada and even predates the Maple Leaf as the symbol of Canada.

Fig. 5.7 Aquatic nutria mammal. (Photograph courtesy of Christopher F. Frazier)

The North American species was introduced to northern Europe in 1937 in response to the near extinction of the smaller, more sporadically distributed native beaver, *Castor fiber*. However, efforts are now underway to remove the North American species from Europe because it tends to overwhelm its smaller Eurasian cousin. Thanks to dedicated and coordinated conservation efforts by wildlife societies and European government agencies, the native Eurasian species has now been reestablished in much of its former range except for Portugal, Italy, and the southern Balkans. It now occurs widely in western Siberia and can even be found in small populations in wooded mountains of northern Mongolia.

Among the significant anatomical adaptations of all beavers are a streamlined body with a massive skull, powerful chewing muscles, 4 chisel-shaped and continually growing incisors for cutting wood (20 teeth in total), somewhat upwardly displaced eyes and ears that are ideally located when the body is mostly submerged, and a tail that is strong, broad, thick, flattened, and suitable as a rudder. The anatomy of the beaver brain suggests that these semiaquatic vertebrates have a relatively high level of intelligence.

Fig. 5.8 North American beaver (*Castor canadensis*) in northwest Colorado. (Photograph courtesy of Flickr contributor Dona Hilkey)

Keystone Species

Beavers are considered "keystone species" in nature because of their profound environmental impacts on the structure of the aquatic and nearby riparian habitats where they alter streams into ponds, thereby changing conditions for the permanent aquatic (fish, amphibians, invertebrates, etc.) and transitory, semiaquatic fauna (waterfowl) along with the nature of the surrounding fully terrestrial landscape and the animals inhabiting those niches. These semiaquatic mammals directly shift the aquatic landscape from lotic (streams) to lentic (wetlands, ponds, and lakes), thereby significantly altering available niches for many other animals. The tree removal and flooding resulting from construction of dams and lodges might initially seem to have only negative effects on the surrounding terrestrial habitat by thinning the adjacent forests. However, changes in the resulting terrestrial and waterside flora over time can positively influence some terrestrial mammals, birds, and other vertebrates who benefit from a greater diversity of flora and reduced density of trees near the previously densely forested stream banks. Beaver dams can also have a number of positive effects on the environment. For example, they can contribute to wetland preservation and restoration, recharge of groundwaters, easing downstream erosion and sediment transport, and decreasing loss of vital nitrogen and phosphorus from the aquatic system. They also create havens for fish, amphibians, birds, and many other species.

Family Life from Kits to Maturity

A pair of adult male and female beavers form a monogamous couple and live within their lodge (described in the next section) with their current and sometimes previous offspring. Family units in a single lake may build multiple lodges if a single one does not suffice. Each year a mother beaver has one litter of one to four babies, or "kits," with this year's offspring joining both the parents and the kits born the previous year in the beaver lodge. Mothers nurse their young for 2–3 months to supplement the bark and leaves provided to them within a week of their birth. As they grow and gain confidence, the kits begin to explore the waters surrounding the lodge, with initial expeditions often involving the young following or even clinging to the tail of an older relative. These expeditions, as with most beaver activity outside the lodges, occur most often (but not exclusively) during twilight or before dawn. By the time they attain the ripe old age of 1 year, the kits join in the processes of dam and lodge construction and repair and often also participate in raising the next year's kits. Although they may permanently leave their parents' lodge after 2 years to venture out on their own, sexual maturity can be delayed as long as 2–3 years after birth. When environmental conditions are challenging from food shortages, they may remain with their parents for an extra year. Even when they permanently depart the lodge, their kinship can be ascertained and the former family member is identified and tolerated based on recognition scents produced by their anal secretions. Beavers establish territories and exclude invaders using in part boundary scent mounds constructed of mud and local debris, augmented by a yellowish urine-based substance (castoreum) from the beaver's pelvic castor sacs.

The typical diet of a beaver consists through much of the year of terrestrial ferns, grasses, woody herbaceous vegetation, various roots, and tree leaves, along with submerged and emergent aquatic vegetation including catttails, rushes, sedges, and water lilies. They commonly retain a stockpile of their preferred tree and shrub species within their lodge and/or in underwater caches in case they are forced to switch to a diet of the bark and cambium by thick ice on the pond's surface. They remain active in winter without hibernating and can swim under the ice for up to 15 min before requiring a breath of fresh air from their lodge or surrounding atmosphere.

Beavers as Carpenters and Ecosystem Engineers

Rather than employing hammers and saws and without access to architectural drawings (except within their genes), beavers have evolved as significant ecosystem engineers by employing their anatomical tools—including dexterous front paws and strong and sharp teeth—to harvest nearby trees and other vegetation for use in building dams and lodges. The dams, as described below (Fig. 5.9), are constructed to retain shallow stream waters and thereby alter them into deeper and safer (for the beaver) ponds and lakes while providing a "real estate lot" in which to build a lodge

Fig. 5.9 Beaver dam on a small section of the Snake River. (Photograph courtesy of Flickr contributor Adams at the National Park Service)

home. The building materials for dams and lodges include tree trunks and limbs, other semiaquatic and terrestrial vegetation, rocks, and mud (in lieu of cement!). Lodges are constructed as year-round shelters (Fig. 5.10) for the beaver and its family, for the capture and storage of food (especially through the sometimes harsh and long winters), and to some extent as sites for communication with their families and perhaps other animals. To accomplish the last goal, a beaver can "speak" with a variety of churs (sharp or trilling sounds), snorts, and whines and can communicate alarm by slapping its tail.

Lodges provide a safe and cozy environment for beaver families, but they often cannot be constructed in deeper or faster flowing water bodies, and instead the beaver may build a retreat within the bank. Such "bank lodges" have multiple passages and may be covered on the exterior with limbs and other plant material. For safety of the beaver family, a minimum water depth of roughly ~2 ft/0.6 m is required to keep the entrance to an aquatic-located lodge underwater, but the ideal depth is closer to 5–6 ft (1.5–1.8 m). If the depth approaches that lower level, the beaver will either try to build a dam or dig a burrow in the bank. To start construction of the lodge, the water level either has to be at least at a minimal depth, or it will need to be increased by building a partial diversion dam. The next step in lodge construction is to form an outer base with wood inserted into the muddy bottom. This is followed by constructing the remainder of the structure with branches and logs dragged from shore and plastered with bark, smaller sticks, and other types of vegetation, all

Fig. 5.10 Beaver lodge. (Photograph courtesy of Flickr contributor Courtney Celley at the USFWS)

"cemented" together with mud. A vent is built into the upper portion of the lodge to maintain suitable internal temperatures in the warmer months.

Tree boles and limbs provide the strength for lodges but are more typical of building materials for dams. A relatively large tree—say nearly 6 in/15 cm in diameter—can be felled in just under an hour, while larger ones can take 4 or more hours. The beaver drags or carries the wood in its jaws. The rocks and mud needed to meld everything together are transported tucked under the chin while being secured with the front legs. Larger trees are often felled just for access to the smaller branches rather than for the trunk itself. Bark on the trunk is removed then or later as a food source for the beaver family.

Beavers living in a stream usually require construction of a dam for protection and associated food sources, while those residing in lakes with minimal outflow generally do not need to construct a dam unless the lake is very shallow. While the dams provide the proper safe conditions for many lodges, they also afford substantial protection from bears, coyotes, and wolves, at least while the beaver is in the water. Dams range in size from perhaps 10 ft/3 m wide to gargantuan structures stretching over 2,140 ft/652 m with a height of 14 ft /4.3 m and a thickness of 23 ft/7 m, as was found near the Montana City of Three Forks. Such dams benefit not only the beaver and provide a safer lentic habitat for aquatic species upstream of the dam, but they also create habitat on the exterior of the dam for small mammals, amphibians, birds, and other animals.

The Beaver's Near Demise and Current Challenges

Prior to the colonization of North America by Europeans, the beaver's natural enemies were primarily bears, cougars, coyotes, and wolves. Native Americans also hunted them for their fur and meat, but overall beaver population was not threatened by this subsistence trapping. Their near demise as a species started, however, with the exploitation of their thick furry skin as an ideal and highly fashionable "fabric" for human coats and hats. The pelts of three or more beavers were required to construct a single hat. So popular were beaver skins in the clothing industry that several wars (such as the French and Indian War of 1754–1763) were partially started as a result of competition in the beaver trade! The biggest actor in this commercial exploitation and species decline was the Hudson Bay Company. For example, during a 10-year period in the mid-1800s, this and other fur companies were purchasing from trappers more than 150,000 beaver pelts per year! Fortunately, the increasing use of synthetic insulation in clothes eliminated the need for warm beaver pelts. Moreover, conservation and animal rights organizations have made that practice no longer profitable as well as bad for a company's image.

With the reduction and finally the near cessation of fur trapping for the clothing industry, you might perhaps conclude that beavers are now safe to return to their natural ecological roles. Unfortunately, such has not entirely been the case. Beavers continue to be killed by individual humans and governments because their dams can lead to the swamping of nearby roads and crop fields and even significant portions of towns. These fascinating and unique animals continue to be slaughtered and threatened as a species because they and their habitat are not protected under the federal Endangered Species Act or by various state laws. For example, in late 2020, the Oregon Fish and Wildlife Commission unfortunately rejected a petition to provide protection for beavers throughout the state. All that stands between them and extinction may be the action of citizens and wildlife organizations to protect them, including with public outcries when the beavers are unnecessarily threatened. Clearly, however, their interactions with communities will require judicious relocation of the beavers in many cases to suitable habitats where they can continue their fascinating natural activities unimpeded by humans.

Chapter 6
Strangers in a Strange Aquatic Land?

What Is So Strange About a Stranger?

Some aquatic species whose names are well-known to most people from electronic or print media are simultaneously considered strangers when they occur within unexpected habitats. Most of these "strangers" in freshwaters would likely include species we normally associate with the ocean. Oddly enough, these seemingly misplaced animals may in some cases have actually migrated to freshwater millennia before *Homo sapiens* trod the Earth—so from one perspective, "we" are more appropriately labeled as strangers in a strange (terrestrial) land.

If You See a Mammal Swimming Next to You, It Might Not Be a Human

Freshwater Seals

Many mammals exploit freshwater environments for drinking water, food, temperature control, and sometimes longer-term habitat; but rarely are they permanent residents, and even fewer are representatives of predominately marine taxa. Among the rarest of the last groups are freshwater seals, including the Lake Baikal seal, *Pusa sibirica* (Fig. 6.1), found in Siberia. For those of you who may be somewhat unfamiliar with Baikal, this Russian lake is the deepest (>1 mile/9.7 km deep) and is the second oldest lake on Earth at 25–30 myr. For comparison, the modern Laurentian Great Lakes of North America developed their current form only about 10,000 years ago. Lake Baikal contains roughly 20% of the world's unfrozen freshwater and is connected by rivers flowing north to the Arctic Ocean while also receiving river water from its southern neighbor Mongolia (Fig. 6.2). This freshwater seal in Baikal

Fig. 6.1 Baikal seal, or "nerpa" (*Pusa si*birica). (Photograph courtesy of Flickr contributor Sergey Gabdurakhmanov)

is very abundant with an estimated population of 60–100,000 individuals. For non-Russians interested in seeing these seals, you may wish to hop a train from Tokyo, Japan, south to the Shiga Prefecture to view the two live Baikal seals kept in their spectacular aquarium. That facility is actually focused on the biology of the 4,000,000-year-old Lake Biwa, one of the few really ancient lakes of the world. While the only true freshwater seal species is limited to Lake Baikal, that Japanese aquarium holds a few seal populations from other, predominately marine species that have subpopulations in lakes located near marine habitats in several countries, including in the Hudson Bay region of Canada. The Caspian seal (*Pusa caspica*) is another seal that occurs in inland waters; but in this case, it lives in this very large but slightly saline lake (about 12.8 parts per thousand sea salt versus 35 ppt on average in the ocean).

Another equally rare mammalian occupant of freshwaters—which I admit is actually a terrestrial mammal rather than an aquatic one—is the Japanese macaque or snow monkey (*Macaca fuscata*; Fig. 6.3). This is a curious species both because it is the northernmost, nonhuman primate on Earth and because snow monkeys primarily exploit thermal spring lakes for their warmth in the winter—which is perhaps a lame excuse for mentioning this curious species in book where it would otherwise not belong!

Fig. 6.2 Map showing location of Lake Baikal in Russia where the Baikal seal lives. (Map courtesy of Wikipedia contributor Kmusser)

Freshwater Dolphins

Dolphins suddenly became a popular marine mammal starting in 1964 with the television program "Flipper." So enamored was the public with this intelligent animal that marine seafood marketers even changed the common name of the completely unrelated dolphinfish (or "dorado" in Spanish) to the Hawaiian name "mahi-mahi" (meaning "very strong"), presumably to avoid a potentially negative association and subsequent loss of sales. Being an active scuba diver, I have had multiple opportunities to dive with these mammals and even observed the training of similar dolphins in the ocean at Anthony's Key Resort on the Honduran island of Roatan in the Caribbean. It is also common for passengers on dive boats and other marine vessels to see dolphins swimming alongside the boat, often catching a free ride on the boat's bow wave. However, while marine dolphins have become very

Fig. 6.3 Snow monkey, or Japanese macaque (*Macaca fuscata*). (Photograph courtesy of Flickr contributor SF Brit)

popular, their freshwater relatives have largely remained unknown outside of the world's tropical and subtropical latitudes.

Freshwaters in tropical and subtropical regions of the world are now home to five species of river dolphins and porpoises, some of which are highly threatened. Two genera occur in South America, with the grayish-pink freshwater dolphin or boto (*Inia geoffrensis*; Fig. 6.4) being relatively abundant in the huge Amazon and Orinoco River basins. This mammal can reach over 8 ft/2.4 m in length and weigh 220 lb/100 kg. Another smaller dolphin species (*Sotalia fluviatilis*) occupies waters of the upper Amazon Basin in Peru. While in Manaus, Brazil, visiting a scientific colleague, I had the delightful opportunity to swim in the large Rio Negro with these grayish-pink freshwater dolphins (Fig. 6.5), even though that required starting from a somewhat flimsy, floating platform that was truly overloaded with fellow tourists. Looking at these dolphins close-up, it was somewhat hard for me to imagine that their closest surviving, evolutionary relatives are hippos (see Chap. 4)!

A few other species of freshwater dolphins occur in southern Asia. The Ganges and Indus River dolphins (*Platanista gangetica*) live in rivers on the border of India where they hunt for prey mostly by employing the ultrasonic vibrations that are common to all dolphins and porpoises. This technique is especially vital when hunting in turbid waters and/or for those species with very poor eyesight. Several rivers in Southeast Asia, including the Mekong, are home to the extremely rare Irrawaddy dolphin (*Orcaella brevirostris*). It is primarily confined to freshwater and brackish rivers and lakes of southern India and the large Songkhla Lake in the Malaysian

Fig. 6.4 Pinkish freshwater dolphin (*Inia geoffrensis*) from the Amazon River of South America. (Photograph courtesy of Flickr contributor Ana Claudia Jatahy)

Fig. 6.5 Tourist attraction on Rio Negro River near Manaus showing the star of the show—a pink river dolphin. (Photograph courtesy of James H. Thorp)

Peninsula of Thailand. Irrawaddy dolphins are known to squirt streams of water up to a distance of nearly 5 ft/1.5 m in order to herd potential prey fish in a desired direction for later capture. Finally, the giant Yangtze River of China once supported two species of these aquatic mammals: the "Baiji" (*Lipotes vexillifer*)—which has not been seen in half a century—along with a "critically endangered" population of the Yangtze finless porpoise (*Neophocaena asiaeorientalis*; Fig. 6.6). The latter apparently still survives despite the constant threats of pollution, reduced food supplies, and bodily damage from ship movements.

River dolphins in general are not widely distributed worldwide, at least in modern times, mostly as a consequence of river pollution and intentional and accidental destruction from humans. The five extant freshwater species differ somewhat in shape and other body features and vary from looking very similar to their marine relatives to being starkly different (such as the Amazon dolphin; Fig. 6.4). For example, freshwater species have no need for the thick insulating blubber that is common to those living in the ocean because their tropical, freshwater habitats are warmer than in the ocean. However, like their marine relatives, they use sonar to locate objects and also communicate with clicks and whistles. Their eyesight is adequate above water, but they appear to have little if any sense of smell. In contrast to their marine relatives, freshwater dolphins and porpoises do not adapt well to captivity and often die within months of their capture.

Sirens of the Tropical Forest: Manatees

All three manatee species (and possibly more subspecies) enter fresh or estuarine waters to some degree, but only the Amazonian manatee (*Trichechus inunguis*) lives exclusively in freshwaters (Fig. 6.7). It was rated by the IUCN Red List for 2021 as being "vulnerable." More recent reports have revealed that some populations of the African manatee (*T. senegalensis*) occur far inland in the interior wetlands of the Niger River, some 1500 aquatic miles (2414 km) from the Atlantic Ocean where most other populations of this species reside. Other African populations have been reported from ocean-connected, tropical lakes in Cameroon, Chad, and Ghana and from many West African rivers.

A species related to manatees is the dugong (*Dugong dugon*). It occurs primarily in the Indo-West Pacific, including the east coast of Africa and at least formerly in the Red Sea. However, it apparently does not reside within freshwaters. It is the closest living relative to the extinct but related marine sirenian—the Steller's sea cow (*Hydrodamalis gigas*) which was discovered by Europeans in the mid-eighteenth century. Sadly enough, all members of that species were killed for their flesh and driven to extinction in less than three decades—a sad record of human ignorance and cruelty. This is especially surprising because other marine species of manatees were supposedly mistaken at a distance for the enticing female mermaids of Greek mythology by possibly drunk and amorous sailors—which is even stranger when you consider that the oldest living relatives of manatees are elephants!

Fig. 6.6 Critically endangered finless porpoise (*Neophocaena asiaorientalis*) found in the Yangtze River of China. (Photograph courtesy of Wikimedia Commons contributor Huangdan2060)

The life of the gentle manatee may seem rather dull, as they spend most of the hours in a day resting on the bottom and only rise to the surface periodically to breathe or feed. Amazonian manatees are herbivores that primarily consume aquatic

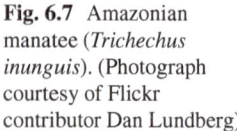

Fig. 6.7 Amazonian
manatee (*Trichechus
inunguis*). (Photograph
courtesy of Flickr
contributor Dan Lundberg)

vegetation (such as water hyacinth and water lettuce) along with any of the abundant ripe fruit that periodically falls into the water. Plant food is collected using the manatee's muscular lips and stubby head bristles, sometimes augmented with use of front flippers. Individual manatees in Florida have also (but rarely) been observed protruding their heads out of water to feed on terrestrial plants. Aquatic plants are macerated using "cheek teeth." These teeth gradually wear down as they move forward in the jaw and are replaced continually through their life by new teeth erupting from the rear, much in the way this process works in modern and ancestral elephants. The teeth of manatees are not effective as antipredator structures.

The life of a freshwater manatee is thought to span at least 30 years when undisturbed and may last up to a decade longer. Reproduction begins around the age of 5 and at a rate of about once every 2–5 years, generally with one calf born each time. This process usually occurs sometime during the mid-December to mid-May rainy season when the forest is flooded. As mammals, manatee mothers nurse their young regularly for about a year and then periodically for the next 12–18 months as the calf is gradually weaned. Adults are generally solitary animals once the young finally depart but may also live in small groups.

These river sirens are reasonably smart mammals with similarities in intelligence to marine dolphins, including long-term memory and some associative learning. They have well-developed communication skills, as clearly shown by the frequent conversations between a mother and her calf.

These mammals, especially those exposed in shallow water, are subject to predation by sharks, crocodiles, and jaguars, but their deadliest enemy is *Homo sapiens*. Some human subsistence hunters kill them for their blubber, skin, and especially meat, most of which are sold at relatively high prices in local markets. A more much serious problem, however, is the accelerating destruction of the rainforest for farming, metal extraction, and oil production. Today some enlightened efforts are being made in Brazil to protect and reestablish manatees in areas where they have been extirpated. However, when visiting an aquarium facility in Manaus, I gazed fondly into a huge outdoor tank which contained perhaps half-a-dozen juvenile manatees that had been rescued after their mothers were killed for their meat by native hunters. A government goal is to reintroduce adults of this mammal species to parts of their original range where they have been extirpated. Orphan calves, like those that had been rescued by zoo/aquarium personnel in Manaus, are moved to relatively safe natural areas once they are large enough to survive on their own.

Freshwater Sharks and Stingrays

Sharks

Before exploring some aspects of sharks found in inland waters, you need to be aware that some "bony fishes" living in freshwater habitats are erroneously called "freshwater sharks," including ones you may see at your local pet store. However, those members of minnow, carp, and catfish families are totally unrelated to true sharks, the latter of which have cartilaginous rather than bony skeletons, among other major differences. Another issue is the bad reputation sharks have in the public's mind. Yes, they are top predators, and yes they are known to attack and sometimes kill humans, especially in shallow coastal and estuarine waters. However, only eight human deaths from a shark occurred from the beginning of 2010 to the end of 2019 in the USA. The exaggerated and largely unrealistic fear of sharks may have started with the 1975 blockbuster movie "Jaws." In contrast, the conservation society "Greenpeace International" estimates that people slaughter 100 million marine sharks each year, and a Miami newspaper calculated in a summer 2022 issue that at least 73 million of these are killed for "shark fin soup," which is considered to be an Asian culinary delicacy. Even if these estimates are far too high, there is no doubt that some of the really vicious creatures of our planet are terrestrial and walk on two legs—or to quote a Pogo cartoon from the 1970s by Walt Kelley: "we have met the enemy and he is us!" In fact, sharks are an extremely valuable top predator in coral reefs and other marine systems and undoubtedly play (or previously played) an important role in tropical freshwater rivers. While we really do not understand the full significance of their ecological role in freshwaters, we should at least have an open mind about this ecologically valuable predator because we know that their depletion on offshore coral reefs is severely damaging them and the same may be happening in freshwaters.

Reports from several continents indicate that several species of normally marine sharks travel long distances up rivers, sometimes posing extreme danger to local humans (Fig. 6.8). Especially hazardous is the species *Carcharhinus leucas*, which is known in different parts of the world as a bull shark, Zambezi shark, and Lake Nicaragua shark. Most adults apparently do not swim very far upriver from their more common ocean habitat, but at least two were caught in the last century by fishermen near St. Louis, Missouri, in the USA—about 600 miles (965 km) from the Gulf of Mexico! A subset of scuba divers around the world (including a then younger and more foolish author of this book) have been known to seek out sharks

Fig. 6.8 Small bull shark (*Carcharhinus leucas*) being measured. (Photograph courtesy of Flickr contributor from the Florida Fish and Wildlife Service)

on marine scuba diving trips. This would not be wise in all cases in rivers or other murky waters, but there are places on the Pacific Coast of Costa Rica where you can take a short boat trip to a tiny offshore marine island to scuba dive with large bull sharks. However, this same species has killed people swimming across the mouth of a nearby river just south of that island. In at least the experience of the author, these sharks stay just at the edge of visibility and follow you around, while all divers in the group are simultaneously consuming inordinate amounts of air from their scuba tanks!

While both sharks and rays (fish class Chondrichthyes) almost exclusively live in the ocean, only three to four living species of true sharks in the genus *Glyphis* predominately inhabit freshwater systems, along with some populations in nearby marine coastal regions. They are most commonly found in Southeast Asia, on the island of New Guinea, and in parts of Australia and India. Only the critically endangered Ganges River shark of India is restricted to freshwaters or low salinity tidal areas for both reproduction and normal life activities. It seems mostly to inhabit the middle to lower section of large rivers. This same shark species may also occur under a different but erroneous name in Pakistan and several other Southeast Asian countries. Threats to this shark species relate to overfishing and general habitat degradation, including from dam construction.

Stingrays

Although marine sharks entering freshwaters are responsible for more human deaths, in fact more injuries result annually from freshwater stingrays (Fig. 6.9), including several species of *Potamotrygon*. Like sharks, this almost entirely marine

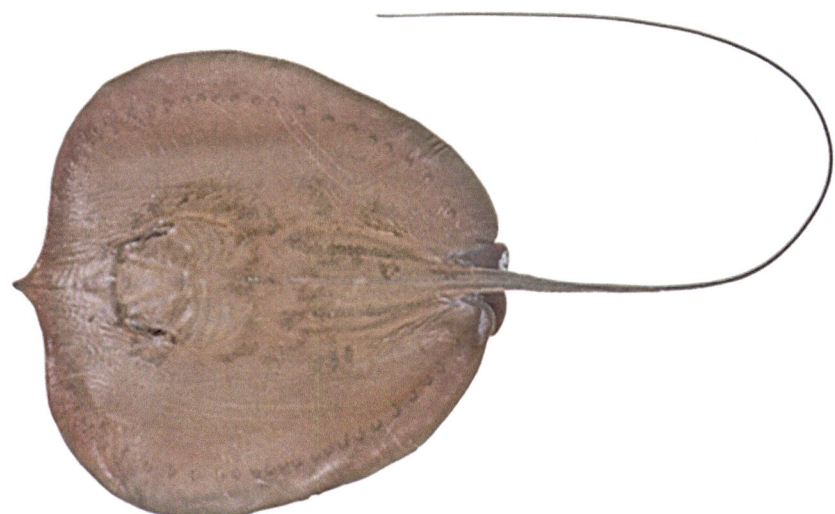

Fig. 6.9 Freshwater whipray/stingray (*Himantura dalyensis*). (Photograph courtesy of the Australian CSIRO's (Commonwealth Scientific and Industrial Research Organization) National Fish Collection)

group of rays and skates are in the vertebrate class Chondrichthyes and thus have cartilage rather than bone in their skeletons. Also like their shark relatives, they use urea in their body fluids not only as a primary waste product but also to balance and regulate the movement of ions into and out of their blood. Unlike sharks, however, their negative interactions with humans relate not to the extremely rare intentional attacks but instead from humans inadvertently grabbing or stepping on the ray and being "stung" by the ray's entirely defensive structure. The best way to prevent stepping on stingrays is to shuffle your feet rather than taking big steps when walking through shallow freshwaters and the surf areas of coastal beaches.

Freshwater stingrays can be found in rivers of four continents (Africa, Asia, Australia, and South America), though in some cases they seem to have been introduced by humans. For example, the large reservoirs in Singapore have been infested with stingrays apparently originating from South America which may have been released by aquarium pet owners who may have become tired of their increasingly large aquatic pets. Past introductions have required extensive and continuing efforts to extract them from the popular reservoirs in this small Southeast Asian country.

While most stingrays reach only 1.6 ft/0.5 m in disc width, the whiptail stingrays (family Dasyatidae; Fig. 6.9)—which have the widest distribution worldwide—are known to reach more than 6.5 ft/2 m in body width in the Mekong River. Stingrays consume a variety of benthic organisms including clams, mussels, various shrimp and crab species, larger benthic insects, and occasionally small fish, worms, etc. They live 5–10 years in captivity, but their natural lifespan in the wild is unknown.

Invaders on the Move

While most of the freshwater animals discussed above migrated eons ago, species invasions are still ongoing, with many aided intentionally or accidentally by humans. [Note: for purposes of this chapter, I am not distinguishing between species naturally dispersing to new environments and those that have been intentionally or accidentally transported by humans from distant lands and waters.] Most of the attempted invasions are unsuccessful, of course, but by trying, trying, and trying again over the years, some species have succeeded against long odds, especially when humans lend a hand. Some native species are frequent, and often successful, short-distance colonizers. For example, some crayfish in the southeastern USA can move among adjacent ponds and wetlands or from one river basin to another if the distance is not too great and if it is humid enough at night so that their gills stay wet (needed for effective respiration) during the migration.

At least two tropical freshwater fish species that can crawl on land between water bodies have invaded the southern parts of North America. Walking catfish (*Clarias batrachus*) is a species of omnivorous freshwater fish native to Southeast Asia that has colonized much of the US state of Florida since the mid-1960s when it was imported for fish aquaculture. By using their stiff pectoral fins, these average size catfish (1.6 ft/0.5 m long) can wiggle along the forest or grassland floor to a new aquatic habitat on a damp evening. Snakeheads (*Channa* species) are another species of fish that has recently invaded multiple states in the USA from another continent, probably through the aquarium trade. They regularly consume fish, frogs, and sometimes rodents and are prized as food by humans in many parts of the world. They are known to migrate among water bodies over distances of up to a quarter mile on moist land at night or under cloudy and humid conditions by wriggling their bodies and thrusting with their fins!

Another human-aided invasion occurred toward the end of the last century when zebra mussels (*Dreissena polymorpha*; Fig. 6.10) and quagga mussels (*D. bugensis*) were introduced from their native Ukraine probably via western Europe and likely inside the ballast water of oceangoing ships that dumped their contaminated ballast water into the North American Great Lakes to lighten the boat's weight in anticipation of replacing the enclosed water with commercial cargo. As described in Chap. 2, these small mussels live secured with strong byssal threads in large clumps on rocks and native mussels (eventually starving or smothering them) as well as on metal surfaces like boats and inside the pipes of power plants, commercial factories, and city water plants. One unique feature that has allowed their expansion is that they produce planktonic larvae or "veligers" that are distributed by water currents, in contrast to the much slower form of fish-aided dispersal of larvae (glochidia) by unionid mussels native to North America. Another invasive species—the Asian clam (*Corbicula fluminea*; Fig. 6.11)—is now present in many of the continent's

Fig. 6.10 Zebra mussel (*Dreissena polymorpha*), an invasive species in the USA that was derived from Europe. (Photograph courtesy of D. Jude at the NOAA Great Lakes Environmental Research Laboratory)

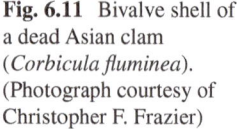

Fig. 6.11 Bivalve shell of a dead Asian clam (*Corbicula fluminea*). (Photograph courtesy of Christopher F. Frazier)

rivers. While it can also clog pipes, it does not have the many byssal threads that allow dreissenid mussels to attach firmly to each other and to other hard surfaces.

Interestingly enough, some aquatic invertebrates frequently colonize distant habitats via air transport by winds or waterfowl. For example, the cysts (desiccation-resistant, fertilized eggs) of fairy shrimps and many other small invertebrates like water fleas (Cladocera) can be distributed hundreds to thousands of miles by wind along with dust particles or inside of or externally on migrating waterfowl (see Chap. 2). The really international travelers, however, are microscopic rotifers whose propagules are known to travel among continents and across oceans using pervasive wind currents.

Chapter 7
Lakes: Turning of the Seasons

Now that you have been exposed in previous chapters to some principal character-istics of freshwater fauna, let's investigate some important aspects of the physical, chemical, and thermal environments where freshwater animals thrive or in some cases barely survive. The information contained within should help you understand the limits faced by the animals discussed in previous chapters. The current chapter focuses on so-called relatively permanent "lentic" habitats—those with minimal outflows, including mostly ponds and lakes of all sizes. Subsequent chapters will cover running water habitats, cave streams, wetlands, and some unique habitats like bogs.

How to Make a Lake

It is easy to take your local lake, natural pond, and wetlands for granted and assume they have always existed. In fact, for most people, natural "lentic" systems—that is, those with either no outflow or where the resident pool is huge compared to the daily outflow (like the North American Great Lakes)—have been present for their entire lives, except for the local ephemeral wetlands. Even most artificial reservoirs in North America other than farm ponds are at least a half-century old. While some natural lakes date back millions of years, most lakes in the northern half of the USA, Canada, and Europe are only a bit older than 10,000 years—the most recent period when glaciers excavated natural basins in many parts of the northern hemisphere.

Lakes, ponds, and ephemeral wetlands can be formed in many ways other than just by glacial scouring and human activities. Movements of the Earth's tectonic plates (which may cause earthquakes) have usually formed the planet's deepest and oldest "rift" lakes. For example, the most ancient of such tectonic lakes is Baikal (Fig. 7.1), which is located in Russian Siberia close to its southern border with Mongolia. Not only is it one of the world's oldest lakes at 25–30 million years, but

J. H. Thorp, *The Otter and the Fairy Shrimp*,
https://doi.org/10.1007/978-3-031-64029-2_7

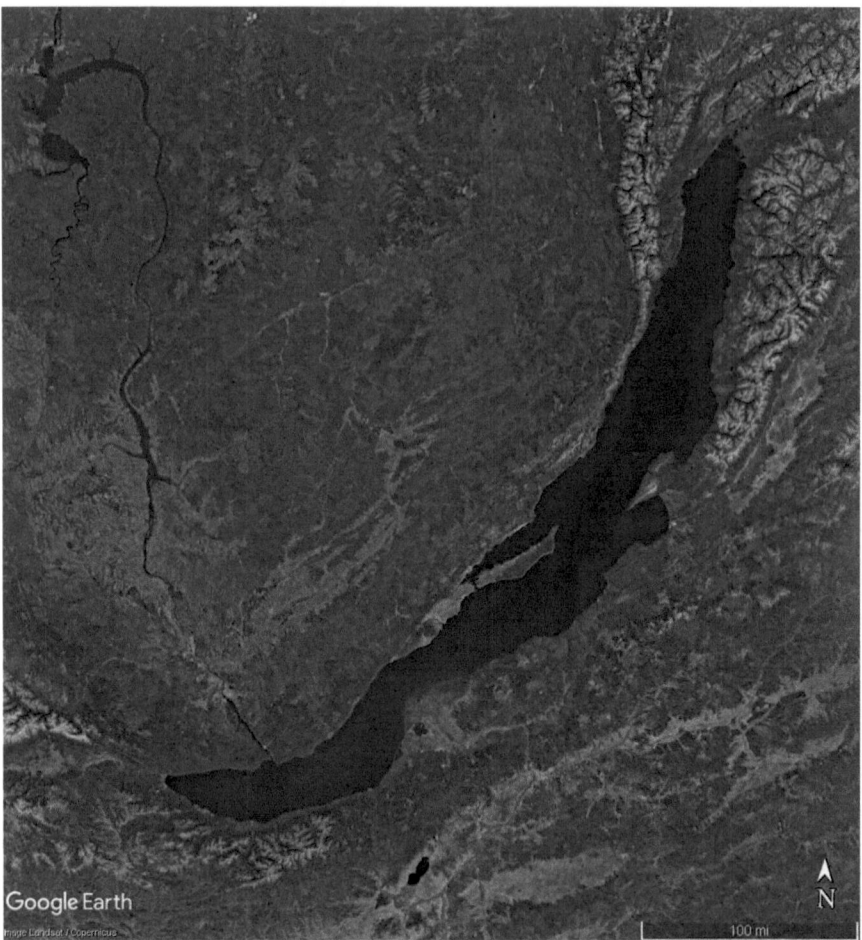

Fig. 7.1 Google Earth Landsat image of Lake Baikal in eastern Russia

it is also the Earth's deepest lake at nearly 5400 ft/1646 m and contains the greatest volume of freshwater at 5700 cubic miles (23,759 cubic kilometers)! Actually, the inland Caspian Sea between Europe and Asia contains more water, but its average salinity is about one-third of that found in the world's oceans. Other giant rift lakes include some in Africa such as Lakes Tanganyika, Victoria, and Malawi. The oldest lake in the world is thought to be Lake Zaysan in the Central Asian country of Kazakhstan, which may have formed 65 myr ago. Lakes can also form within volcanic calderas, such as Crater Lake in Oregon, USA (Fig. 7.2), with the resulting lake generally being very deep, steep-sided, and clear. One biological characteristic of ancient lakes is that they often feature species found nowhere else in the world, including the Baikal seal (Fig. 7.3; see Chap. 6) along with other unique species of fish and invertebrates.

Fig. 7.2 Crater Lake, a volcanic lake in the US state of Oregon. (Photograph courtesy of Flickr contributor John Manard)

Fig. 7.3 Seals from Lake Baikal in Siberia. (Photograph courtesy of Wikipedia contributor Nina Zhavoronkova)

Fig. 7.4 Century-old photograph of an ephemeral pool or "buffalo wallow" made by American bison. (Photograph courtesy of Willard Drake Johnson (photographer) and the US Geological Survey)

"Lentic" systems can also develop from chemical processes (solution lakes), landslides, wind scouring (aeolian lakes), meteorites, and the movement of river channels that isolate lateral water bodies (fluvial lakes). And as was discussed earlier in this book, animals can also form short-lived lakes and pools, including through intentional construction by beavers and incidental formation from ground rubbing by bison ("buffalo wallows") and various large African mammals (Fig. 7.4). Many of these very shallow to very deep lakes contain a suite of both specialized and generalized organisms. To better understand these habitats and the organisms living there, let's start by exploring the physical nature of mostly semipermanent lakes and ponds to understand how they contribute to the abundance and diversity of aquatic organisms while sometimes limiting their distribution.

Ice Rises

The ice in your cold drink floats … "so what does have to do with lakes?" you might ask. Well, if that were not the case, you would probably not be reading this book— indeed we along with most freshwater animals may never have evolved on Earth or at least lived anywhere where water freezes in the winter! Don't believe it? Well, think back to a hot sweltering summer day when as a kid you dived into a warm pond to cool off, only to come to the shocking realization that the pond was almost icy cold at a depth of perhaps 10 ft/3 m! Is there a link between these two observations? From a scientific perspective, the answer is unquestionably "yes."

During the summer in the temperate zones on Earth, the temperature at the bottom of lakes and ponds that are deep enough to avoid complete turnover is often around 39.2 °F/4 °C because that is the temperature where water reaches its maximum density and thus sinks to the bottom. At the lake's surface, the temperature could easily exceed 90 °F/32.2 °C—a level where the resulting water's oxygen level can be low enough to kill fish because warm water holds far less oxygen than cold water. In contrast, during the winter, the bottom of any lake that does not freeze solid may hover around 39.2 °F/4 °C. But near the surface, the water—now frozen—has reached its lowest density at 32 °F/0 °C and moves upward in the pond … "ta-da" …, thus, ice floats in ponds and in your glass of ice water or a coke from the local burger joint!

What does this have to do with life on Earth? Well, if the maximum density of ice were at 32°F rather than the minimum density, then all but the shallowest of ponds could be frozen from the bottom to near the top even in the summer, much like our remaining glaciers survive the summer heat. The biotic effects would have been that multicellular life might not have appeared in rivers and lakes and, who knows, perhaps later evolved into small land animals and eventually into readers of this book!

Turn, Turn, Turn!

The layering of lake water described above and the seasonal changes in water temperature have profound effects on aquatic life that includes a short period of mixing and lake turnover. In the summer, two distinct vertical sections of water are present: an upper warm layer (which scientists call the "epilimnion") and a lower cold layer known as the "hypolimnion." These major areas are separated by a thin area of rapid temperature change known as the "metalimnion." The upper layer contains true algae as a food base and enough oxygen to support many types of fish and invertebrates. Vital dissolved nutrients (nitrogen and phosphorus) continually decrease as the summer progresses toward autumn, thereby limiting growth of true algae. In contrast, the lake's bottom layer often features low to zero levels of oxygen, little-to-no light, relatively high amounts of nutrients, some normally toxic compounds, and a large diversity of bacteria-like organisms. The latter organisms include those that can transform methane gas into products usable by single-cell protozoa. In the winter, the same layers exist, but they are less prominent because of much smaller differences in temperature and water density. In the autumn and spring, however, the lake very briefly features the same temperature and water density from top to bottom. At that time, any significant wind can cause the lake to "turn over," mixing the former static water layers and redistributing nutrients. Lake turnover commonly occurs once (in "monomictic" lakes) or twice (in "dimictic" lakes) per year in the temperate zone, with the former mixing in the winter and the latter taking place in the autumn and spring. Algal blooms follow each turnover, especially in the spring because of the influx of nutrients previously locked within the bottom layer. This lake turnover is critical to yearly lake productivity and producing the fish important for kids with cane poles as well as adults competing in televised bass fishing competitions.

Two exceptions to this phenomenon are worth noting. In "meromictic" lakes, the lower portion of the bottom layer is slightly to very saline and possesses a higher water density. This results either from natural seepage into the lake of compounds from the underlying bottom or surrounding watershed or artificially from industrial release of salty water from manufacturing plants. In either case, the entire lake never turns over completely, and nutrients are partially trapped near the bottom. In really deep lakes, like some of the Laurentian Great Lakes shared by to the USA and Canada (Fig. 7.5), so much water is present that seasonal winds are unable to turn over the entire lake. In contrast, in very shallow ponds, including ephemeral playa wetlands (Fig. 7.6) that support fairy and tadpole shrimp, the water depth may not exceed 3–7 ft (0.9–2.1 m), and the entire pool may mix on a daily basis.

In contrast, rivers typically do not experience this same phenomenon because the water is continually being redistributed by turbulent flows (small streams to some rivers) or by a vertical, helical flow pattern (like a stretched out "slinky toy") from the surface to the bottom as the water moves downstream (see Chap. 8).

Fig. 7.5 MODIS (terra) satellite image of the North America Great Lakes. (Photograph courtesy of Flickr contributor NOAA Great Lakes CoastWatch)

Fig. 7.6 Playa lakes in western New Mexico, USA. (Photograph courtesy of Flickr contributor formulanone)

Life on the Edge

Lakes and rivers teem with life, but much of it is clustered near shore. Such areas often enable small organisms to locate food and gain protection from larger ones by seeking shelter in havens provided by rooted aquatic plants (Fig. 7.7) and sometimes wood snags in this littoral zone. Wood snags along the borders of lakes and rivers teem with attached and motile invertebrates, often in greater diversity in comparison with areas with substrates of only mud, sand, or gravel. Fish—especially young ones and adults of smaller species like various minnows—thrive in this "littoral zone" as do many invertebrates such as crayfish, mussels, dragonflies, and sometimes leeches. Invertebrates may find shelter from predators by burying themselves in soft bottom sediments, hiding under rocks, or clinging to the camouflage of aquatic plants. Food is relatively abundant and diverse in the littoral zone compared to mid-lake habitats. However, turbulence from waves can be deadly in this nearshore area for aquatic species, with the constant threat of being washed ashore or thrown against rocks. Moreover, wading birds (Fig. 7.8) may pluck unwary fish or aquatic insect from their supposedly safe shelters among aquatic vegetation.

In contrast, those invertebrates living in the middle of lakes—which scientists call the "pelagic zone"—tend to be very small and mostly transparent, thereby making it more difficult for fish to capture them in clear waters. These minute creatures (usually <0.04 in/0.1 cm long) most often consist of copepods (Fig. 7.9a), cladocera (Fig. 7.9b), rotifers (Fig. 7.9c), and sometimes carnivorous phantom midges

Fig. 7.7 Littoral zone of a lake showing aquatic vegetation. (Photograph courtesy of Flickr contributor Ivan Radic)

Fig. 7.8 Wading bird in a lake. (Photograph courtesy of Christopher F. Frazier)

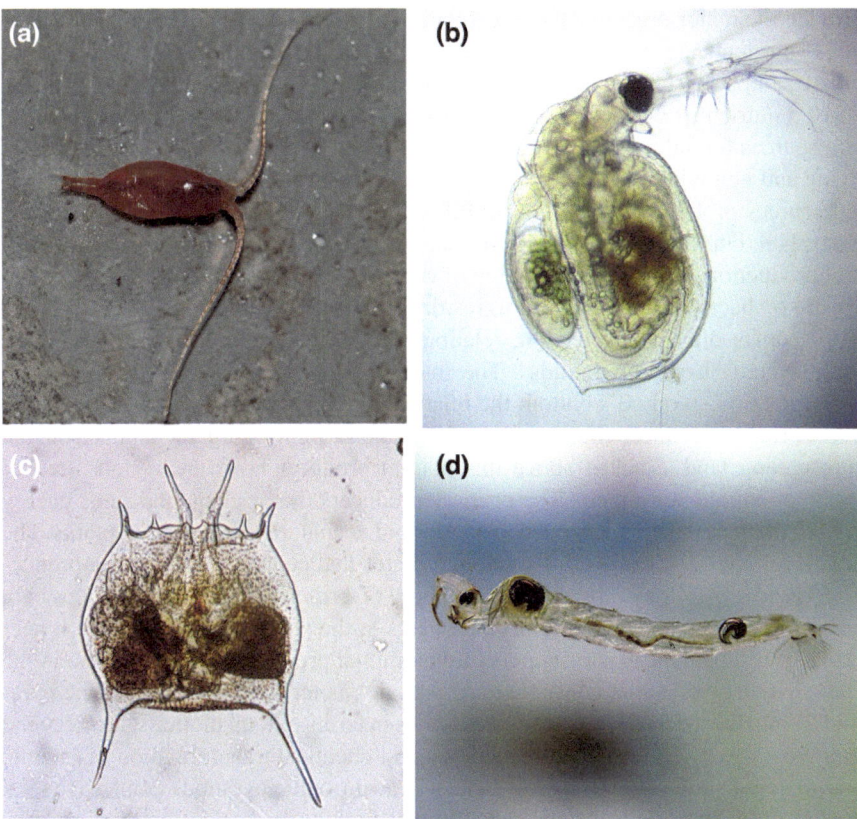

Fig. 7.9 Photographs of (**a**) a calanoid copepod, (**b**) the cladoceran *Ceriodaphnia dubia*, (**c**) the rotifer *Brachionus quadridentatus*, and (**d**) the phantom midge *Chaoborus*. (Photographs courtesy of iNaturalist contributors kbkash (photo "a") and Ken Kneidel (photos "b and c") and Wikipedia contributor Piet Spaans (photo "d"))

(Fig. 7.9d), and they are preyed upon by fish that either grab them individually or use their gills to sieve the prey from the water. True algae (phytoplankton) and photosynthesizing cyanobacteria (previously called blue-green algae) are abundant in the pelagic zone down to a level of 1% of the surface light. Photosynthesis is not possible for true algae below that light level, but photosynthetic bacteria may continue photosynthesizing in areas somewhat deeper in the lake.

Food Webs

Scientists that study lakes and wetlands (= limnologists in the strict sense) still debate the sources of energy that support populations of invertebrates and fish in lakes. Some argue that aquatic-generated nutrients from lake primary producers provide the major organic energy to support animal food webs. These "autochthonous" nutrients are derived mostly from algae and to a lesser extent from aquatic mosses and both rooted and floating vascular plants. Others argue that terrestrially derived nutrients (= "allochthonous" carbon) from land plants are the most important sources for animal growth and maintenance and general system metabolism of living and nonliving (abiotic) processes in lakes. The balance of internal and external sources probably varies a lot in different types of lakes and their surrounding watershed, but the weight of evidence seems to favor aquatic sources for lake animal production. However, production of carbon dioxide in lakes may be supported mostly by bacterial processing of terrestrial grasses, leaves, etc.

Scientists often study feeding relationships in lakes and rivers by examining multilayered "trophic pyramids." The traditional pyramid can be constructed by showing at the pyramid's bottom the relatively large amount of biomass or energy in the primary producer compartment of the lake along with organic inputs from the surrounding land. Stacked above this bottom producer layer are various smaller levels of animal consumers. The primary producers that package the sun's energy consist of algae, autotrophic cyanobacteria, and nearshore and terrestrial plants. The next level up the pyramid consists of herbivores that consume the primary producers. These are followed by smaller numbers of primary, secondary, and possibly tertiary predators until the primary producers at the pyramid's base can no longer provide sufficient energy to support further animal production above it. Also in the lake are detritivores that consume dead organic matter from all levels of the pyramid. While most detritivores are invertebrates in terms of total biomass, they depend to a great extent on the initial breakdown and chemical transformation of organic matter by bacteria. Although most human-drawn trophic pyramids emphasize total annual production and develop a traditional looking pyramid, some approaches focus on existing biomass (= standing stock biomass), in which case the pyramid may even appear inverted.

Some ecologists are interested in what controls the number of levels of the pyramid and the total biomass in each level. Other scientists examine feeding pathways and predator-prey relationships in terms of who eats who and what affects those relationships. While there are many ways of determining feeding relationships, including dissecting the "gut contents" and carefully watching predators eat, the most exact approach in most cases is to trace certain elements through the web using chemical techniques, such as stable isotope analyses of carbon, nitrogen, or sometimes sulfur.

Those Nasty Cyanobacteria (Blue-Green Algae)

In an increasing number of reservoirs and some natural lakes (such as Lake Erie), autotrophic cyanobacteria (previously and erroneously called blue-green algae; Fig. 7.10a, b) are becoming overwhelmingly important in the summer. This primarily occurs as a result of excessive nutrient inputs from municipalities and

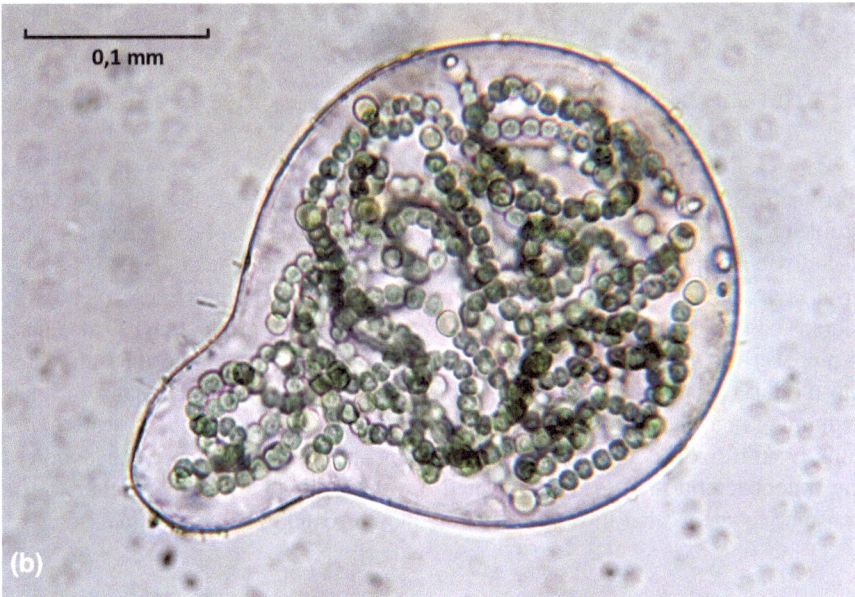

Fig. 7.10 (**a**) Bloom of cyanobacteria ("blue-green algae") in a lake and (**b**) a microscopic view of an *Anabaena* cyanobacteria colony. Photographs courtesy of (**a**) Theodore Harris at the University of Kansas' Kansas Biological Survey and Center for Ecological Research and (**b**) Flickr contributor Phillippe Garcelon

Fig. 7.11 Lunker bass killed from the effects of a cyanobacteria bloom in a lake. (Photograph courtesy of Flickr contributor at the Florida Fish and Wildlife agency)

surrounding farmland. When the abundance of cyanobacteria is low, environmental problems are minimal; but when their production is great, some of them can produce toxic compounds that directly kill fish (Fig. 7.11) and invertebrates and may produce "taste and odor" problems in municipal water supplies. Excessively large blooms can severely reduce the light energy available for production of true algae. A more serious event may occur when the cyanobacteria die in large amounts, causing the lake to become anoxic and killing any remaining fish. Humans and pets swimming in a lake with a cyanobacteria bloom may also be sickened by this exposure. Even if toxic cyanobacteria are not abundant, the total primary production of the cyanobacteria may lead to depleted oxygen at night when the cyanobacteria consume oxygen rather than producing it, and this can kill aquatic animals.

Protecting Our Aquatic Systems

Aquatic systems provide an abundance of valuable ecosystem services to the environment and humans, but only if they are allowed to function in a normal, safe manner. For the long-term protection of freshwater organisms and to ensure a steady supply of freshwater water for human consumption and recreational activities, it is clear that we need to do a better job protecting our water bodies and their surrounding watersheds. Environmental groups within federal, state, province, municipal, and indigenous nations have the task of setting and enforcing vital environmental standards. These are greatly aided by the many local to international nonprofit, environmental and wildlife organizations that are especially adept at observing nature and informing agencies and governments of developing environmental problems. Not-to-be-forgotten is the limited number of commercial companies that regularly set aside funds for environmental health in their role as good citizens. Readers of this book can aid those government, commercial, and private organizations by raising your voices when you detect developing environmental problems.

Chapter 8
Still Waters Run Deep

Of People and Rivers

The history of human civilization is intimately tied to flowing waters, especially permanent rivers because they have historically been large enough to provide people with dependable supplies of water, food (mostly fish), transportation, and power generation from dams. They also add nutrients to adjacent, lowland farmlands when rivers periodically flood. To confirm this historic relationship, take a look at a globe or world map on the web and search for old cities that have an important role in human history, and then see how many are perched on a river bank. Some prominent cities fitting this bill are Baghdad (Tigris River), Cairo (Nile R.), London (Thames R.), Paris (Seine R.), and Shanghai (Yangtze R.). Some historically more recent cities near rivers in North America are Albuquerque (Rio Grande), Kansas City (Missouri R.), Montreal (St. Lawrence R.), New Orleans (Mississippi R.), and New York City (Hudson R.), to name just a few. In addition to these medium-to-large cities, there are many small-to-large towns present on six continents that are adjacent to streams capable of supplying adequate water to meet the needs of local people. Where rivers are not sufficiently large and plentiful, large lakes and coastal oceans typically provide some of those resources. In probably the worst case, underground water reservoirs can be tapped for needed water. Of course, despite the many services rivers provide, they can also cause serious problems for us that are principally related to periodic flooding and occasionally diseases carried by aquatic organisms like mosquitoes (Diptera, Culicoidea) and blackflies (Diptera, Simulidae), both of which are more abundant in still and slowly flowing waters.

J. H. Thorp, *The Otter and the Fairy Shrimp*,
https://doi.org/10.1007/978-3-031-64029-2_8

Physical Attributes for Defining Rivers: Power Function, Turbulence, and Flow

Most rivers originate as ephemeral headwater creeks and progressively enlarge as they flow downstream to the ocean, while others arise from the outflow of lakes (e.g., part of the St. Lawrence River of North America) or large wetlands (e.g., Paraná River of South America). Other rivers terminate in lakes or wetlands. In a few cases, rivers occupy basins where flow to the ocean is not possible. These "endorheic" or "terminal basin" rivers may simply evaporate downstream, disappear permanently underground, or flow into an isolated lake, some of which are saline lakes. An example of the former is the Carson River (Fig. 8.1a, b) that starts in the snows of the high Sierras of eastern California and passes downstream into Nevada where it evaporates in a shrub desert. An example of the latter is the Bear River of Utah which flows north barely into Idaho and then turns sharply south and eventually drains into the Great Salt Lake of Utah (Fig. 8.2). The original lake was the 1,000 ft/305 m deep Lake Bonneville, but a breach in the river's sandstone wall in southern Idaho allowed inflowing river water to largely escape leading the remaining downstream lake water to gradually evaporate to its present salty state. The Bear River's access to the Pacific Ocean via the Snake River was blocked thousands of years ago by regional tectonic activities.

Given that the terms "creeks, streams, and rivers" provide little information about their actual size, scientists have adopted somewhat quantitative terms to describe the magnitude of stream flow within a watershed through use of the "stream order" technique (Fig. 8.3). This approach and more precise metrics allow a rough comparison of biota and stream health among similar-sized streams. Using this classification scheme, the smallest permanent stream is designated as stream order #1. When two #1 streams join, the larger stream is labeled as stream order #2. Two #2 streams joining produce a #3, but adding a stream of a smaller size to one of a larger size (such as adding a #3 to a #4) does not increase the number. Generally speaking, stream orders #1–3 are considered headwaters, #4–6 are small to mid-order rivers, and #7–12 (or maybe 14) are labeled as large or even great rivers (Fig. 8.4). The Mississippi River near New Orleans is considered a "great river" and attains a stream order of #12 by some accounting, while the Amazon River may attain a stream order of #14. Keep in mind, however, that a river with a stream order of #5 may be smaller or larger than a #5 river in a basin farther away where the annual precipitation is different.

Rivers of all sizes exhibit some turbulence in flow. This type of water flow generally decreases downstream, and it is most visible to people at waterfalls, or in shallow rocky rivers, or in headwaters. Turbulent flow benefits fish and other animals by increasing the oxygen in the water, but it can also dislodge attached invertebrates and wash them downstream or into the mouth of a trout or other fish (see next section). Knowledge of turbulence and stream flow are important to people wading in streams because the power of flowing waters can be very threatening. One of the more dangerous things a camper can do is to place a tent near a gently flowing

Fig. 8.1 Terminal basin river: (**a**) upland, high energy portion of the Carson River in Nevada, USA; (**b**) lower portion of the Carson just before it reaches an evaporation basin. (Photographs courtesy of J.H. Thorp)

stream in a canyon or narrow valley with nearby, steeply sloped sides and then go to sleep with confidence that all is well. Some campers have lost their equipment or even lives when an unseen or unappreciated rainfall event upstream caused the

Fig. 8.2 Bear River of Utah where it enters the Great Salt Lake. (Photograph courtesy of Sandra Uecker at the US Fish and Wildlife Service)

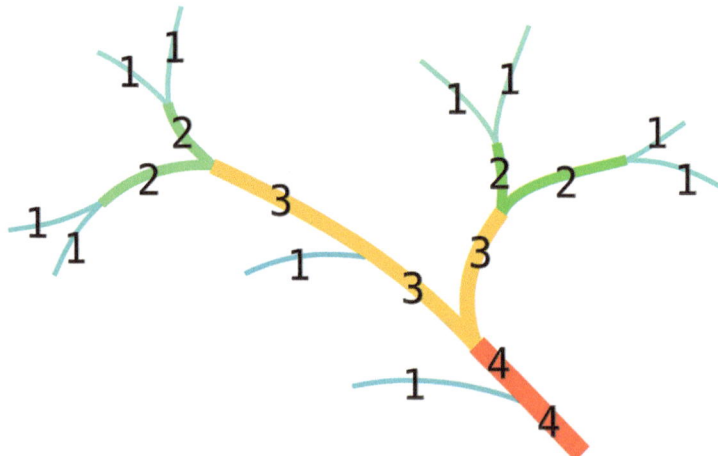

Fig. 8.3 Figure showing how streams in a river basin are given numerical designations. (Photograph courtesy of Wikipedia contributor Kilom691)

adjacent stream flow to rise rapidly and sweep through their campsite. It is also important to understand that the lack of obvious turbulence downstream does not imply the absence of danger. The old saying that "still waters run deep" is true for large rivers. It is also very easy to underestimate the flow rate and power of the river,

Fig. 8.4 Ohio River (considered a large river) above Louisville, Kentucky. Also shown is a tourist steamboat. (Photograph courtesy of James H. Thorp)

both of which usually tend to increase progressively downstream toward the ocean. You can observe this in a river boat by throwing an attached anchor off the bow (never the stern if you want to avoid sinking your boat) and watching the change in the level and turbulence of the water as it hits the sides of the boat.

Most sections of a river are subject to over-bank flooding. In general, as a river's size increases, the lateral extent of the flood waters expands, but fortunately people have more warnings of the impending flow because the river level rises more slowly downstream, and you gain advance warnings of what is happening upstream. If the floods are typically driven by snowmelt or monsoon rainfall, they are generally more predictable in timing than those driven by aperiodic rainfall. Floods in the north temperate zone tend to be briefer and less predictable than those in tropical areas that are impelled by monsoon downfalls. For example, the floods in the basins of the Amazon and Congo rivers may last almost half a year and occur at nearly the same time every year. The longer and more predictable are those floods, the greater possibility that life history aspects and annual production of invertebrates, fish, other vertebrates, and floodplain vegetation will be intimately coordinated with the seasonal timing of these floods. As humans, we naturally tend to assign negative connotations to the term "floods," but in fact flooding is a critically important aspect of the life of a river and the surrounding watershed as described partially in the next section. For this reason, nutrients from the nearby river make adjacent "bottom lands" highly fertile for farming in many years. Research in the scientific field of "flood ecology" is just beginning to emerge as a recognized research topic among aquatic ecologists.

The Good and Bad of Dams and Levees

Humans modify rivers in many ways, such as by extracting clean water, adding wastewater, releasing chemical and thermal pollutants, catching fish, and altering the physical structure of rivers. The last occurs when engineers build dams, construct levees (Fig. 8.5), remove wood snags near the banks, and add shoreline riprap of rock or various artificial material. Riprap sections are constructed to prevent erosion or costly lateral movement of the river channel and to maintain desired depths for barges and ships (Fig. 8.6).

High and low dams are constructed for a variety of reasons important for humans, including electrical generation, flood control, navigation, and recreation. High dams can be defined as those greater than about 50 ft/15 m tall. The largest high dam in the world is the Three Gorges Dam on the Yangtze River in China (Fig. 8.7); it is 1.4 miles (1.6 km) from bank to bank and 630 ft/192 m tall! Construction of high dams is still going at a fast pace in some countries, especially in Brazil, China, and India, but the last high dam in the USA was finished in 1979 (on the Stanislaus River in California). The primary reason for ceasing construction of large dams in the USA was that dams have been clearly shown to have very detrimental environmental impacts on aquatic life and ultimately the functioning of the ecosystem. Low-head dams, that is, those less than 50 ft tall, are even more common in North America because they provide electricity to small communities and some water reserves. In earlier centuries, many dams were used to power watermills to grind grain into flour. In some rivers, such as the Upper Mississippi, seasonal or relatively permanent dams enable the transit of ships and barges during low water conditions via locks

Fig. 8.5 Maintenance road on a levee near Vicksburg, Mississippi, USA. At this time, flood waters of the Yazoo River had reached all the way up to the levee road. (Photograph courtesy of Unsplash+ contributor Justin Wilkens)

Fig. 8.6 Barges traveling on a large river with lateral levees. (Photograph courtesy of Flickr contributor Paul Schultz)

Fig. 8.7 Three Gorges Dam on the Yangtze River in China (shown in the background), which is the largest dam in the world. (Photograph courtesy of James H. Thorp)

Fig. 8.8 Lock on the Upper Mississippi River. (Photograph courtesy of Flickr contributor Tony Webster)

(Fig. 8.8) when shallow river depths approach ~9 ft/2.7 m in depth. In some other rivers, they enable passage of tourist steamboats and commercial barges over shallow rapids, such as the Falls of the Ohio River at Louisville, KY, between the shores of the states of Kentucky and Indiana.

In the natural state of large rivers, the presence of multiple channels is very common. To reduce the periodic effects of floods on the surrounding cities and croplands and to provide a minimum water depth for barge traffic on large rivers, governments have often built levees to constrain the river to the main channel. While such levees have been very beneficial in preventing most of the floods that threaten people and property, they have also produced extensive damage to aquatic life by eliminating the normally highly productive side channels (Fig. 8.9) where many fish and invertebrates lived in relative safety from the high flow conditions of the main channel. These areas are sometimes called "slackwaters" or more commonly "backwaters," though the latter term strictly speaking refers to areas that lack through-flow for part of the year. Levees also increase current velocities and decrease important ecological interactions with the adjacent floodplains. Surprisingly enough, the overall effects on humans from construction of levees are sometimes negative because the adjacent "bottom lands" in the floodplains no longer receive the rich nutrients deposited previously during floods that promoted high crop yields. More startling in some situations is that the economic cost to humans can be higher after levee construction than in pre-levee conditions because people sometimes move on to the floodplain thinking they are safe. But if the levees and flood walls

Fig. 8.9 Tertiary side channel of the Upper Mississippi River. (Photograph courtesy of James H. Thorp)

collapse, the insurance cost in constant dollars (i.e., adjusted for inflation) can be higher than before the levees were built! This happened in the 1990s on the Mississippi River during repeated 500-year floods. From natural history and recreational perspectives, access to these slackwaters (Fig. 8.9) is also important in enhancing overall animal production, and they provide safe recreational havens for houseboats (Fig. 8.10) and fishing boats that are often not available in the river's main channel.

The Shape of Rivers Now and Then

Ask most people to draw the shape of a river, and they will probably show it with meandering channels, perhaps with an island or two thrown in for artistic reasons, if nothing else. While a meandering shape is common, especially in flatter terrains, sections of rivers have different shapes at various points from upstream to downstream. This condition is a response to what we call hydrogeomorphic forces and is influenced by valley characteristics, channel slope, substrate size, and sometimes terrestrial vegetation. The nature of the river from upstream to downstream is not always entirely predictable, though distributary shapes (much like the roots of a tree) are more common near the terminus of the river, and braided channels are more common in shallow headwaters or even mid-order to large rivers like the

Fig. 8.10 Large houseboat attached to the shore in a side channel of a large river. (Photograph courtesy of Flickr contributor flowcomm)

Tagliamento River of southern Europe (Fig. 8.11). Multichannel systems like "anastomosing and anabranching" river sections tend to be found in the lower half of rivers, as in the Mississippi, and may form temporarily isolated pools, as in an oxbow lake in North America or a billabong along Australia's Murray River (Fig. 8.12). These are characterized by multiple, interconnected channels on alluvial plains with important diversity of current velocities, organisms, and ecological processes. Rivers of the Great Plains are typically sandy and shallow and sometimes benefit from use of airboats (Fig. 8.13) to sample extensive areas on them. However, these propeller-driven airboats are more commonly used in snag-filled areas of the backwaters of floodplain rivers. Constricted channels appear most commonly in mountainous regions, such as in the Grand Canyon of the Colorado River (Fig. 8.14) and the Chishui River of China (Fig. 8.15); but while those are more commonly present upstream, they can also occur closer to the ocean.

In Mark Twain's river stories and books (particularly *Life on the Mississippi*), rivers had "reaches" and "bends" the exact location of which the astute river pilot was forced to commit to memory and periodically update as channel conditions changed. A reach was originally defined as a long stretch of a relatively straight channel—much like an extended arm when reaching for something—while a bend is like the shape of your arm when you flex it at the elbow. Water moves faster in the outer part of the bend (think of it as your so-called funny bone), while the inner part of the bend is shallower and may contain a sandbar. The deepest part of the river follows a generally winding path (called a thalweg) as the water moves from side to

Fig. 8.11 Braided channel in low flow conditions in the Tagliamento River of northeastern Italy. (Photograph courtesy of Flickr contributor Alessio Milan)

Fig. 8.12 Billabong (= oxbow lake) on Australia's lower Murray River. (Photograph courtesy of James H. Thorp)

Fig. 8.13 A relatively complex airboat useful for working in shallow rivers (like the sand-bottomed Kansas River shown here). (Photograph courtesy of James H. Thorp)

Fig. 8.14 Constricted channel of the Colorado River in North America. (Photograph courtesy of Flickr contributor Didier Camus)

Fig. 8.15 Constricted channel of the Chishui River in southwestern Chaina, a tributary of the Yangtze River. Also shown is a pagoda along this river. (Photograph courtesy of James H. Thorp)

side and also turns over continually on its path to the sea, much like the shape of a child's stretched out slinky toy. Steamboats generally traveled upstream closer to the shore where the currents were slower but stayed in the middle of the channel when going downstream to exploit faster currents and to avoid the wood snags present near shore (Fig. 8.16) that could breach the hull and sink a wooden boat. Interesting enough, the US federal government supported development of "de-snagging" boats in the mid-1800s to remove these dangerous logs from the shoreline of navigable rivers. Unfortunately, removal of the shoreline snags is detrimental to the fish and invertebrates that need this shelter and hard substrate to avoid predators and to capture the more abundant food found on projecting wood in these areas.

Home in the River: Where Critters Live and Eat

From massive alligator snapping turtles to nearly invisible zooplankton, rivers contain a great diversity of species from their headwaters to their mouth. Mayflies, hellgrammites, caddisflies, scuds (amphipod crustaceans), snails, and even sponges can be found in wadeable headwater streams and along the banks and lateral slackwaters of large rivers, while fishing spiders may be seen skating across the surface of quiet stream pools. [See Chaps. 1, 2, 3, 4, 5 and 6 for more information on and photos of aquatic animals.] Using a kick net and stirring the upstream rocks and

Fig. 8.16 Wood snags along the shoreline of the Upper Mississippi River. (Photograph courtesy of James H. Thorp)

Fig. 8.17 Madtom or stonecat catfish (*Noturus flavus*) often found among stones of clean headwater creeks. (Photograph courtesy of iNaturalis contributor reuvenm)

sediment with your feet in permanent headwater streams are an easy way to collect small fish like rainbow darters and madtom catfishes (careful—these small cats have poison in the pectoral fins; Fig. 8.17). Crayfish are common in smaller streams where potential fish predators are less abundant, but you can also find them in large numbers in medium-to-large rivers where abundant woody shelter is present as refuges. A good way to collect crayfish is to use a minnow trap baited with a piece of chicken or a small can of pet food or tuna (with a few holes punched in it) and placed in the river by the bank and secured to a stationary object on land overnight. [Make sure an aquatic snake has not entered the trap before you open it!] If you venture into the creeks and rivers of the southern hemisphere, crayfish are almost entirely replaced by freshwater crabs and shrimp. Both these other crustaceans occur in the northern hemisphere, but their diversity is low, and crabs generally do not venture far upstream from the estuaries. Actually, there are many more species of shrimp and crabs worldwide than there are species of crayfish.

Freshwater mussels and clams may be locally abundant, especially in the south-eastern USA, and centuries ago were occasionally eaten by indigenous people around the world, despite the fact that they are not as tasty as marine bivalve mollusks. However, you will not see them for sale now in the USA because their flesh is not as savory as their marine cousins. As described in Chap. 2, freshwater mussels occasionally produce pearls, and these can be used in necklaces sold in many jewelry stores (Fig. 8.18) even though their quality is far surpassed by marine pearls. At present, there are concerns about overharvesting freshwater mussels for later use as seed pearls in marine pearls produced in Asia, as this increases the chance of extinction of North American freshwater taxa already threatened by pollution.

As the stream grows in size, the diversity and size of fish increase, and playful otters may be spotted as well as turtles basking in the sun on a log or sandbar. You may also spot a muskrat fishing and possibly an elusive racoon with its paws dipped

Fig. 8.18 Pearly constructs (including pearls) produced by the mantle of freshwater unionid mussels in response to irritation of the mussel's fleshy mantle. (Photograph courtesy of Wikipedia contributor Jennifer Gaglione)

in the water to soften the tissue of its prey in the stream before eating it. If you live in the southeastern USA, watch out for venomous snakes like water moccasins living along the banks or resting on raised stumps in the swamp and periodically seen swimming in the creek or swamp.

In headwaters, the turbulence can easily dislodge aquatic insects and other invertebrates and wash them downstream where a wily trout hiding in the calmer waters present behind a rock may dash from its refuge and snap them up. Many invertebrates hide under the rocks during the day where they are relatively safe from the water currents and predators like rainbow darters (Fig. 8.19), but at night they may crawl out on to the upper rock surface to feed on algae and other food sources. Of course, the turbulence is also present at that time, but predatory fish are less likely to sight their prey at night. Both intentional and accidental "drift" peaks at night. The former may occur because the invertebrate is seeking more food on a downstream rock or is escaping an invertebrate predator. That may raise in your mind the question "How can there be any invertebrates upstream if they keep drifting downstream?". The answer to that question is that some invertebrates (e.g., amphipods or "scuds"; Fig. 8.20) regularly swim upstream, while many adult aquatic insects emerge on land, fly upstream, and then deposit their eggs in the nearby stream. Among the last group are mayflies, which emerge nearly simultaneously in a region—and at times, they may be so abundant (Fig. 8.21) that they can cause skin irritation in people or even lead to traffic accidents by blocking visibility, much like a fog does.

While marine invertebrates and fish are generally far more colorful than freshwater species (aside from the occasional rainbow darter), it can be fascinating to study the diversity and ecology of freshwater invertebrates in streams and lakes. Many states have manuals that list the species of either fish (most frequently) or mussels found in their state or within adjoining localities. These are often written by scientists at state agencies or individual non-government authors and are easy to locate on the web. To learn more about inland water invertebrates, please see one or more of the publications listed at the end of Chap. 2.

Fig. 8.19 Rainbow darter (*Etheostoma caeruleum*), a beautiful fish found among cobble substrate in headwater streams of North America. (Photograph courtesy of Flickr contributor Yankech gary)

Fig. 8.20 Freshwater amphipods (*Gammarus mucronatus*) from the York River in Virginia. (Photograph courtesy of Flickr contributor from the Smithsonian Environmental Research Center)

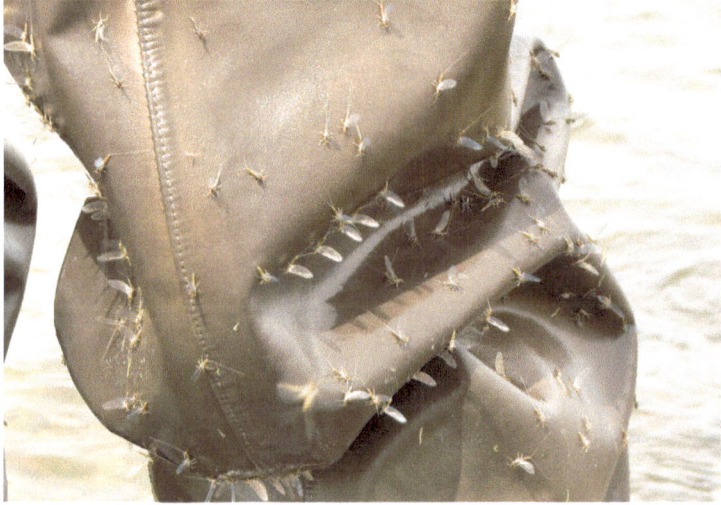

Fig. 8.21 Swarm of nonbiting, adult mayflies (Ephemeroptera) next to the Green River at Seedskadee National Wildlife Refuge. (Photograph courtesy of Flickr contributor Tom Koerner of the US Fish and Wildlife Service)

Chapter 9
The Darkness Underfoot

The Waters Below

This chapter and Chap. 10 will introduce you to some aquatic habitats and their resident organisms that may be less well-known to many readers. The current chapter focuses on the physicochemical and biotic features of cavernous, subterranean environments (usually called caves). I also briefly summarize important aspects of the typically microscopic "interstitial" faunas that are often located between grains of sand and larger rocky particles below the bottom surface of rivers and sometimes among buried gravel present lateral to some streams. But, the primary focus will be on the nature of underground caverns and their scientific study (= speleology). Before venturing into the darkness underfoot, let's first take a brief look at the interstitial fauna of rivers.

Interstitial Habitats

The first thing to understand is that the vast majority of freshwater ecologists rarely study interstitial habitats, and the mostly microscopic species that reside there are generally poorly known compared to taxa in other lotic and lentic habitats. Nonetheless, we classify them into two basic physical zones linked to some degree with rivers: the hyporheic and phreatic zones. The former has a closer physical connection to rivers and is found just under the stream bed and/or closely lateral to a creek or river where access to flowing water is easier. It contains a greater diversity of species and range of sizes than can be found in the phreatic zone—the zone which is farther laterally from the stream and barely influenced by the stream's flow. Invertebrates in the hyporheic zone include many minute crustaceans, such as blind amphipods. Most of the fauna is confined to these habitats for their entire lives, but also present is a low diversity of aquatic insects that need to emerge as flying adults. While food is very limited in these habitats compared to surface streams, two advantages accrued by the residents are that the environment is far more stable and vertebrate predators are absent. The study of these groundwater habitats and their physical and chemical characteristics is principally in the province of geologists and hydrologists.

Other Habitats Where the Sun Doesn't Shine

Now let's turn to subterranean aquatic environments—habitats that are better known to the public and ecologists alike than are hyporheic faunas. For the purpose of this chapter, I will call these underground habitats "caverns" and will define "caves" as a subset of caverns with an external opening large enough for humans, bats, and a few other terrestrial animals to enter (Fig. 9.1). Occasionally the cave opening is spacious enough for temporary occupancy by animals as large as bears, but in most cases the entrance is small. Aquatic species confined to these caverns for their entire lives are called aquatic "troglobites" (= stygobites), while those that can complete their life cycle either inside or outside of caves are known as aquatic "troglophiles"

Fig. 9.1 Entrance to Mammoth Cave in Mammoth Cave National Park in the US state of Kentucky; note the walkway in the lower right. (Photograph courtesy of Flickr contributor Warren LeMay)

(= stygophiles). Finally, "trogloxenes" live in caverns but must exit through a cave opening to complete their life cycles outside of the cavern. If you are really intrigued by these habitats, I suggest that you start by searching on the web for a moderately large glossary of cavern and karst names for a thorough and more precise coverage of such terms. The analysis below focuses on karst environments, that is, those where the environment is formed primarily from carbonate rocks like limestone.

The diversity and abundance of most aquatic creatures primarily reflect the amount of food, the physical complexity of the habitat, and extremes of temperature (heat and cold) in their surroundings. Extremes of temperature and moisture are more often associated with lower animal diversities. Of these habitat characteristics, thermal extremes are generally not high on the list of concerns for subterranean creatures because underground temperatures tend to reflect the yearly average of the surface waters, that is, they are typically neither too hot nor too cold, unless influenced by nearby hot springs. This may come as a surprise to those of you who have toured a commercially operated cave in the summer and found it relatively chilly but come back in the winter, and it will seem quite mild compared to the land surface. Moreover, if you have squeezed through openings in the wall of that cave on a commercial tour, you might consider the habitat to be exceedingly complex; but for aquatic creatures, the opposite is true because the stream bed tends to be solid rock, occasionally cobble (as in Fig. 9.2), or with a thin covering of sand or mud. In both aquatic and terrestrial ecosystems, species diversity is usually closely associated with habitat complexity. Woody material that could potentially wash in from surface

Fig. 9.2 Stream bed in a Brazilian cave showing rocky stream bed below and stalactites above. (Photograph courtesy of Flickr contributor Rosanetur)

streams would provide environmental complexity, but this material is exceedingly rare in caverns. Consequently, the physical and chemical habitats are not easily subdivided by different species, unlike the above-surface habitats such as forests, grasslands, streams, and lakes with their typically varied habitat complexity that provides areas for expanded habitat niches.

Aside from a lack of habitat diversity in caverns, all resident aquatic species face the daunting problem of obtaining scarce food resources. The carbon base of the food web is mostly derived from outside the subterranean environment, and it can enter in only two ways. First, organic matter can flow into the cavern through surface flow, either via a stream or seepage from the bottom of a karst lake. Depending on the nature of the inflowing stream, this could provide the subterranean food web with either dead organic matter (= detritus) colonized by bacteria and fungi (often the most nutritious component for detritivores) or rarely living animals and algae. Any surface animals or algae entering this way are quickly consumed. A second pathway is through bats (Fig. 9.3) flying into a cave, clinging to a cavern wall, and defecating directly into the stream or indirectly on the stream bank, thereby providing a substrate for bacterial growth at the base of the food web. Insects might also fly into the cave and then die there. In both cases, the energy enters the upper levels of the cavern food web through invertebrate or vertebrate scavengers or more likely via bacterial detritivores. Once the initial food item is consumed, it moves a short distance up through the food chain to either a top predatory invertebrate species or in some cases a cavern fish or salamander. A third related pathway involves

Fig. 9.3 Bat in a Madagascar cavern. (Photograph courtesy of Flickr contributor David Dennis)

chemoautotrophic oxidizing bacteria that often use cave gases (hydrogen sulfide) as an energy source rather than sunlight. [Chemoautotrophic oxidizing bacteria are ancient taxa of bacteria that can form complex molecules often by oxidizing sulfur or iron instead of carbon.] This autotrophic (= self-generated from chemicals in this case) pathway is apparently very rare in freshwater caves and is mostly limited to those with sufficient sulfur compounds (= sulfidic caves), but it is well-studied in marine volcanic vent communities on the ocean floor.

Cavern, Cave, and Spring Creatures

As a general rule and in comparison with surface streams, cavern creatures are characterized by a low number of species, few individuals per species, a low density of all taxa present, a unique fauna compared to surface streams, low reproductive rates, and generally exceptionally long lives for troglobitic species compared to some related surface species.

One of the seemingly odd things about these ecosystems is that the density of their resident species is extremely low "within" all caverns, while at the same time, the uniqueness of their fauna "among" isolated caverns is usually very high. This isolation and low population size in a given cavern make the species highly vulnerable to extinction. This is the primary reason that many of their animals are listed on national and state lists of threatened and endangered species. The reason for this apparent anomaly is that the isolation of cavern systems across the landscape promotes evolutionary divergence of species. This occurs because the original colonists have more opportunities to evolve into new species over centuries of isolation. One group of crustaceans—isopods (aquatic sowbugs; Fig. 9.4)—for example, has more unique species in caves than in surface water bodies of North America! This is understandable when you consider that it would be very difficult for a troglobitic species from cavern "A" to exit through the cave opening and travel by surface stream to a separate cavern "B" without being consumed by surface predators even if they could physically reach a distant cave.

In general, troglobitic creatures are characterized by a longer lifespan compared to related surface creatures. In association with their long lives and the cave's low ambient food sources, troglobitic species like blind cave crayfish tend to have a very low reproductive rate. In contrast to surface species, however, they often invest more energy in individual eggs so that offspring can survive long enough to begin acquiring food on their own. Cave-bound critters also tend to be sedentary and slow-moving at best in order to conserve their limited access to food sources. As noted above, some ancient cavern creatures—including various fishes (Fig. 9.5), salamanders, and crayfish that have been confined to the dark environment for eons—have lost the use of their eyes through evolution. The original ocular structure is often visible in some form, but the creature can no longer perceive shapes even when a light is provided by humans.

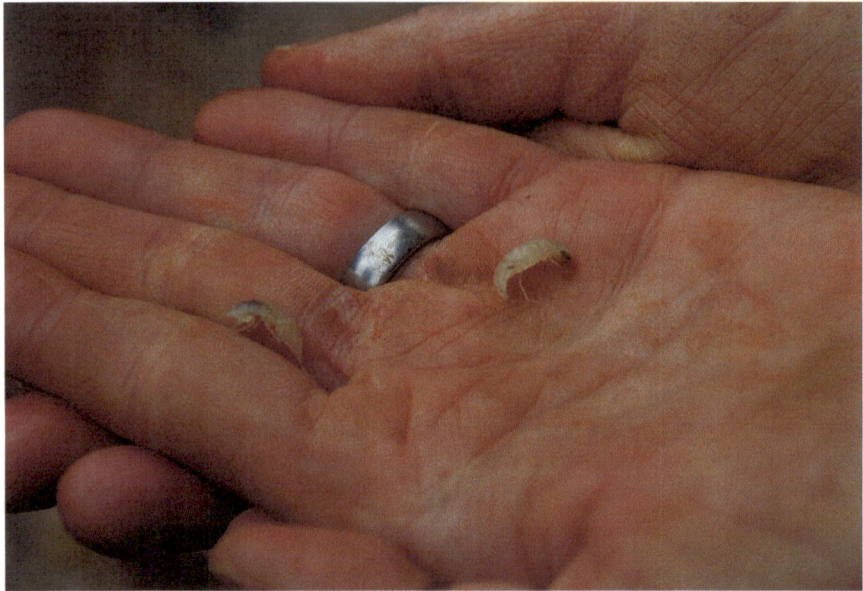

Fig. 9.4 Two isopods from Madison Cave. (Photograph courtesy of the US Fish and Wildlife Service, Northeast Region)

Fig. 9.5 Mexican blind cavefish (*Astyanax mexicanus*). (Photograph taken at the Newport Aquarium in Kentucky and courtesy of Flickr contributor James St. John)

While some surface-dwelling aquatic animals may live weeks to decades (vertebrates like fish, reptiles, and mammals), most invertebrates in streams and lakes live at most a few weeks to rarely more than one year. Exceptions can be found in aquatic habitats closer to the poles where some invertebrate species may live two or rarely more years. In contrast, cave bats may live up to 25 years, while blind cave crayfish and salamanders are known to survive for 50 to even 100 years! The latter is thought to be the maximum age of the blind cave salamander known as "olm" (*Proteus anguinus*) (Fig. 9.6).

If you participate in a commercial cave tour, you are likely to see only bats and maybe some troglophilic crickets, with the former pointed out by the tour guide and the latter usually ignored. There are, however, additional species farther back in the cavern where the tour does not lead, including in rare cases aquatic crickets

Fig. 9.6 Blind cave salamander olm (*Proteus anguinus*) from the Dinaric Alps of Southern Europe. (Photograph courtesy of Flickr contributor Javier Abalos)

(Orthoptera) that can feed underwater before returning to the surface. These are obligate cave species and thus contrast with all other members of this terrestrial order of insects.

If a cave opening is present, there will almost certainly be a greater diversity of species within the cave area where troglophilic and trogloxenic residents can enter and leave the cave more easily. If water exits a cave opening to an outer stream or pool, the outflowing stream will contain troglophilic, trogloxenic, and possibly troglobitic species. However, ecological research in the karst region of southwestern Missouri has shown that those cave species typically disappear within as little as 300–350 ft (91–107 m) downstream. This may reflect a lower competitive ability of troglobitic species compared to normal surface residents, or they may have succumbed to predators, or the cave creatures may not be as tolerant of the abiotic conditions in the surface waters.

If the cave waters exit into a spring lake, the environmental conditions change from a stream-like (lotic) to a lakelike (lentic) environment (Fig. 9.7). One might find troglophilic species here but rarely far into the downstream creek or river. Because of the environmental stability of the subterranean waters, spring lakes and spring creeks often are vital refuges for downstream surface creatures which may be able to swim upstream to a stable aquatic environment that will permit their survival from downstream droughts until better stream conditions reappear. Cave-fed lakes often support a higher species diversity than other nearby natural lakes of similar size but lacking a spring connection.

Fig. 9.7 Spring-fed pool from a karst cave in the US Ozark region. (Photograph courtesy of Flickr contributor Scout House at the Natural Resources unit of the US National Park Service)

Ending This Chapter with a Special and Important Warning to Readers

If you choose to explore caverns, keep in mind that these often-exciting habitats can also be very dangerous places. Even if we assume you can reliably find your way back out of the cavern through a cave opening in most conditions and you do not drown while scuba diving [*underground or deep marine diving can be very dangerous even for experienced divers*], you could still get trapped underground by rising water in a cavern, especially but not exclusively where water flows into rather than out of the cave entrance. The reason this could occur is that by the time you see a rise in the water level far into the cave, the stream level may have sealed you inside by the action of water that has risen to the ceiling of a narrow, low passage closer to the cave entrance. An actual and very unfortunate example of this phenomena made the international news in 2018 when 12 members of a junior football (soccer) team and their coach were trapped in a cave in Thailand for 2 weeks by rising waters just prior to the monsoon season. All 13 individuals were eventually saved as a result of the action of *3000* volunteers and members of their armed service; but as a result of this major operation, one of the rescuers (a former Thai Navy Seal) later died of a blood infection contracted in the cavern. Therefore, if you plan to explore a cavern through an entrance with an inflowing stream, be absolutely certain you have checked the weather radar and know that rain is definitely not in the forecast! Also, make sure someone staying on the surface knows the cave's exact location and when you should return. Note: this is not a problem that you will likely encounter in most commercially operated cave tours—which strongly emphasize safety on their guided tours.

Chapter 10
Wetlands, Bogs, and Some Other Bizarre Habitats

Ephemeral to Permanent, Mostly Lentic Systems of a Sometimes-Unusual Nature!

This book's last chapter is, I admit, a bit of a grab bag because it includes coverage of several kinds of aquatic systems that often receive less attention from scientists and respect from the public than do streams and lakes. Most of this chapter's coverage is on wetlands (ephemeral for months to decades) and acidic bogs, but I also briefly dip my writing pen into thermal streams. A better understanding of the ecological role of wetlands worldwide is vital because we have lost over 20% of their physical coverage since the 1700s, with most of the losses in China, Europe, and the USA (higher percentages have generally been estimated based on extrapolations from certain economically advanced countries).

Defining Wetlands

Thus far, I have focused on describing relatively permanent bodies of water—rivers, lakes, and caves—and the organisms inhabiting them, but not all important aquatic ecosystems are permanent. In fact, some hold water for only a few weeks and yet are critical for the life cycle of many aquatic and terrestrial species. Indeed, as the title of this book implies, some ephemeral, fishless systems hold signature species, like fairy shrimp (Fig. 10.1), and almost all play a vital role for migrating waterfowl and temporary homes for immature and adult individuals of terrestrial vertebrates like frogs and salamanders. Moreover, they contribute to water filtration and groundwater recharge. Depending on the breadth of your definition, wetlands can range in size from a few yards/meters across to thousands of square miles/kilometers, with the latter including areas such as the Amazon River basin; the West Siberian Plain of Russia with its many swamps and floodplains; the tropical flooded grasslands of South America's Pantanal

J. H. Thorp, *The Otter and the Fairy Shrimp*,
https://doi.org/10.1007/978-3-031-64029-2_10

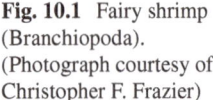

Fig. 10.1 Fairy shrimp
(Branchiopoda).
(Photograph courtesy of
Christopher F. Frazier)

found in parts of Argentina, Brazil, and Paraguay (an area = ~75,000 square miles or ~194,000 km^2); and the freshwater-estuarine mangrove wetland forests of the Sundarbans in the Ganges-Brahmaputra Delta of India and Bangladesh. Wetlands may hold water for more than a year or evaporate and refill several times during a year. They may partially dry while retaining a few isolated pools or remain completely parched for multiple years or even decades while still supporting aquatic life hidden as resistant stages in their sediments until the vital waters return. As an extreme example of the last group, a "clay pan" in western Australia had supposedly not filled for a century; but when sufficient "centennial rains" finally arrived, huge numbers of apparently fertilized eggs (= cysts) of fairy shrimp (Fig. 10.2) successfully hatched out—perhaps some of which may have been there for as long as 100 years! These basins can also be vital water sources for humans and other local mammal species. But on the negative side, they are important sources of mosquitoes and the greenhouse gas methane. Some people reasonably consider alluvial swamps—those with a nearly constant, lateral connection to a river—to be wetlands, but I will exclude coverage of those vital systems in this chapter because they better fit within the definition of the lateral riverscape of a floodplain river (see Chap. 8).

How Wet Are Wetlands?

In a broad sense, inland water wetlands encompass bogs, fens, marshes, peatland mires, playas, pocosins, sinks, swamps, vernal pools, baygalls, and even shallow water-holding wallows hollowed out by large mammals such as bison and elephants.

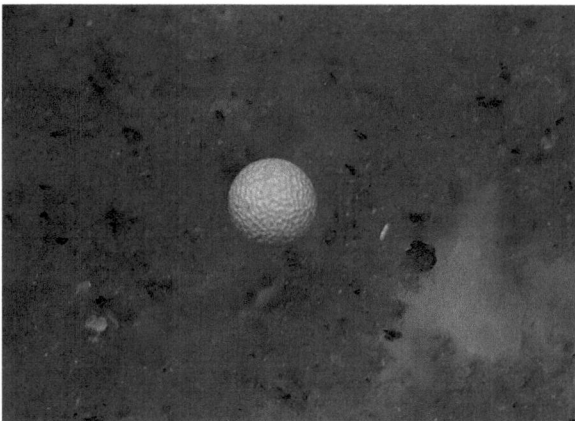

Fig. 10.2 A cyst, which is an environmental resistant reproductive stage of a fairy shrimp, in this case of *Branchinecta cornigera*. (Photograph courtesy of iNaturalist contributor Shawnb2)

And, of course, names for these and other wetlands vary globally even in adjacent countries—for example, a "carr" in Ireland and much of England equates to a "kerr" in northeastern England and Scotland. Most wetlands consist primarily of freshwater, but some are salty, and a few contain naturally toxic compounds, such as borax or even arsenic—all chemicals primarily derived from the surrounding terrestrial basin. The remainder of this chapter's coverage is restricted to ephemeral-to-more-permanent systems that are stand-alone ecosystems and will thus not include details on marshes found along the borders of some rivers and lakes that tend to hold water for much longer periods.

The US government defines "wetlands" as ecosystems with two of the three following characteristics: (1) standing water with water-adapted vegetation, (2) soil that is reliably saturated for all or a significant part of the year, and/or (3) areas that give evidence of typical wetland hydrology. Only the former category supports a significant diversity and abundance of aquatic invertebrates. One or more of these definitions can describe ephemeral wetlands, including playas of the US Great Plains (Fig. 10.3a, b) which may totally lack emergent vegetation when water-filled or gain low-lying vegetation as they mature before drying seasonally to desiccated sediments with or without vegetation. Bottom surfaces of these diverse types of wetlands span a large range from solid rock floor basins, hard dry clay pans, sandy clay soils, silty sand, and anoxic organic "mucky" soil to wet soil supporting plants year-round but with aquatic animals present only seasonally and then with only invertebrate residents. Vegetation in the surrounding land may range from sparse grasses or low-lying woody shrubs (as in prairies) or include overhanging forest trees (as in deciduous forests of the eastern USA).

Fig. 10.3 (**a**) Ephemeral pool in southeastern New Mexico; photograph courtesy of Wiebke Boeing; (**b**) playa wetland in the Great Plains; (Photograph courtesy of Brian O'Neill)

These aquatic systems can be vital as sources for groundwater recharge, as contributed to the Ogallala Aquifer of the US High Plains. The Ogalla Aquifer spans several states in the central USA and sustains the massive agricultural output for which the region is known. That underground system is fed mostly by widespread rainfall on prairie soils as well as water initially contained in small-to-large playas. The latter include, for example, over 20,000 playas in Kansas alone and an even greater number in Texas. Unfortunately, such Great Plains wetlands often occur in prime farming and ranching country and thus frequently face threats to their survival from row-crop plowing and grazing by cattle herds and even semidomesticated bison. The combination of substantial extraction for agriculture and elimination of playas has reduced the natural recharging of these aquifers and has left the Ogallala facing water depletion, leading to calls for playa restoration and more sustainable methods of farming and ranching.

Are Wetlands Different from Ponds and Streams?

Wetlands are lands that hold water for periods from weeks to a few decades but most eventually dry temporarily within a century unless they become shallow lakes. However, this section primarily focuses on temporary wetlands rather than emphasizing those aquatic systems that continuously retain standing water through many decades even though such systems are legitimately classified by governments and scientists as wetlands. The distinction is important because if such systems are more-or-less permanent, they are likely to be deeper for longer periods and possibly may be connected to a stream. This could allow them to support limited populations of several small fish species—and that would dramatically alter the normal community compared to largely fishless wetland systems.

Depending on how you define a river and a wetland, some flowing water systems become marshlike on a yearly basis or even dry completely. Large tropical rivers like the Amazon include huge expanses of wetlands in their lateral floodplains that yearly retain water for months or even permanently, and many of its resident creatures could more accurately be described as wetland animals instead of river organisms. These may represent, therefore, the opposite ecological extremes from wetlands like playas that completely dry at least once per year. The ecologically vital Okavango River of southern Africa has a somewhat intermediate condition, as it dries to a downstream marsh in most years with both isolated and continuously connected bodies of water. A few aquatic vertebrates have life cycles enabling them to survive those drought conditions, such as the very ancient lungfish (Fig. 10.4) found in Africa, Australia, and South America. These fish can easily survive low oxygen conditions in non-flowing water, and many species may even estivate in the dry bankside mud when the stream water totally disappears. This reflects their possession of an unusual, highly specialized and complex respiratory system for a fish that in some ways resembles those of terrestrial vertebrates.

Fig. 10.4 Speckler-bellied lungfish (*Protopterus aethiopicus*) in Japan's Toba Aquarium. (Photograph courtesy of Flickr contributor Joel Abroad)

While the hydrologic cycle of wet and dry periods defines all wetlands, there are other significant seasonal changes as well. These pools can differ extensively in the length of the aquatic period, salinity (total concentration, elemental composition, and potential toxicity), bottom characteristics (some combination of detritus, mud, sand, and/or rock), depth, presence of small to large amounts of submerged and emergent vegetation adapted to hydric soil, and length of the aquatic period. All these factors combine in some degree to influence the abundance and diversity of aquatic organisms.

An Abundance of Wetland Creatures

Probably the best-known wetland creatures in many parts of the world are migratory waterfowl, but other vertebrates and invertebrates are also common in these aquatic habitats. While fish are rare in ephemeral wetlands (unless small, isolated pools survive in large wetlands during the dry season), some wetlands in North America contain populations of very small fish, such as the widely distributed (by humans) "mosquitofish" (*Gambusia affinis* and *G. holbrooki*; Fig. 10.5). These fish are tiny enough to be easy prey to both wading birds and large invertebrates such as dragon-fly nymphs in the insect family Aeshnidae (Fig. 10.6)—the adults of which are sometimes called "hawkers" or "darners." The common name of these small fish is quite applicable, because they regularly eat mosquito larvae (Fig. 10.7) which can be abundant in wetlands and roadside ditches that lack predatory fish. These fish can also survive for long periods in small pools of larger wetlands. Most other fish can-not survive in wetlands unless they can live in those small pools or estivate through the dry period. This is a common dry-season refuge within the Pantanal (Fig. 10.8)—South America's largest wetland which occurs in parts of Bolivia, Brazil, and Paraguay.

Knowledge of wetland waterfowl (Fig. 10.9)—and indeed of wetlands in gen-eral—have been aided by organizations, such as "Ducks Unlimited." While some readers may not approve of hunting migratory birds, such organizations have

Fig. 10.5 A female eastern mosquitofish (*Gambusia holbrooki*) collected from the Rhode River in Maryland, USA. (Photograph courtesy of Robert Aguilar at the Smithsonian Environmental Research Center)

Fig. 10.6 An aeshnid dragonfly nymph (Odonata, Anisoptera). (Photograph courtesy of Flickr contributor Sam Stukel at the Gavins Point National Fish Hatchery in Yankton, South Dakota, USA)

Fig. 10.7 Mosquito larva
(Diptera, Culicidae).
(Photograph courtesy of
Flickr contributor Rob
Cruickshank)

Fig. 10.8 Pantanal waterhole resident crocodiles and giant capybara rodents in South America.
(Photograph courtesy of Flickr contributor Vinícius Mendonça)

provided the political power over the decades to protect wetlands; for example, see
Section 404 in the Clean Water Act of 1972. Wetlands provide vital short-to-longer
layover sites (= bed-and-breakfast habitats?) for migratory ducks, geese, and other
waterfowl in their long flights north and south within and among continents. Many
other nonmigratory species of birds (Fig. 10.10) also use wetlands, such as the lake
proper and wetlands associated with the saline Mono Lake (Fig. 10.11) of eastern
California and areas around the Great Salt Lake of Utah. These can be employed by
the waterfowl as rookeries and as hunting grounds where birds prey on dense

Fig. 10.9 Hooded mergansers in flight. (Photograph courtesy of Christopher F. Frazier)

Fig. 10.10 Winter mallards. (Photograph courtesy of Christopher F. Frazier)

populations of brine flies and fairy shrimp (in particular *Artemia monica*, a species of "brine shrimp"; Fig. 10.12)—organisms that can tolerate waters often saltier than the ocean! Grebes (Fig. 10.13), gulls, and phalaropes are part of the millions of birds from nearly 100 local and migratory species that benefit from the lake and

Fig. 10.11 Mono Lake. (Photograph by James H. Thorp)

Fig. 10.12 Brine shrimp (*Artemia monica*) from Mono Lake in California. (Photograph courtesy of iNaturalist contributor rlawrenz)

Fig. 10.13 Pied bill grebe. (Photograph courtesy of Christopher F. Frazier)

wetlands of Mono Lake alone. Some birds, like the Wilson's Phalarope (Fig. 10.14), molt their flight feathers at Mono Lake, feast on large densities of highly nutritious brine shrimp, and then continue their annual migration to South America once they have fledged a new set of flight feathers.

Other Wetland Residents

While some wetlands are saline, few are significantly or lethally salty; and as a result, most can support a surprising diversity of animal life such as tiny crustaceans, frogs, spadefoot toads (an endangered species), snakes, insects including graceful dragonflies, pesky mosquitoes, permanent and migrating waterfowl, and sometimes small or periodically hibernating fish.

The fauna of wetlands is highly influenced by the amount of water present, how long it lasts, the type and abundance of aquatic and surrounding terrestrial vegetation, and the temporal dependency of individual animals on aquatic habitats. All those characteristics are in turn primarily influenced by the vegetation and climate

Fig. 10.14 Wilson's Phalarope (*Phalaropus tricolor*) at Seedskadee National Wildlife Refuge. (Photography courtesy of Tom Koerner at the US Fish and Wildlife Service)

of the local ecoregion, such as prairies or forested areas. Consequently, the local fauna varies significantly from pools in the moist forested states of the eastern US to species occupying the more open and ephemeral pools of the western states.

Animals basically have three options for life in wetlands (see more on the invertebrates in Chap. 2). One strategy is to hatch, grow rapidly, reproduce within 1–2 weeks, and then produce environmentally resistant, fertilized eggs, thereby allowing the species to survive if the water disappears for months to decades. Examples of these are large branchiopods (including tadpole shrimp; Fig. 10.15, and see Chapter 2 on other taxa), common water fleas like *Daphnia* (Fig. 10.16), copepods, and a few snails, along with various microscopic invertebrates like rotifers and the ever-present, single-cell protozoa (see Chap. 2 for more discussion of freshwater invertebrates). A second strategy is to hatch, grow into adults, depart the pool for short-to-long periods, and then return to lay eggs for the next generation. These include, for example, salamanders (Fig. 10.17), frogs, and spadefoot toads—some of which are included on state and/or federal lists of threatened or endangered species. And third, they can fly into the wetlands, rest and consume energy from local prey for short-to-long periods, and then fly to other wetlands on their migratory path—in one sense, much like all of us might do on a cross-country driving or bicycle trek when visiting hotels, restaurants, and gas pumps or electrical charging stations!

Fig. 10.15 Tadpole shrimp
(*Triops* sp.). (Photograph
courtesy of Christopher
F. Frazier)

Fig. 10.16 Water flea (*Daphnia* sp.) in a drop of water. (Photograph courtesy of Flickr contributor
Brian Tomlinson)

Fig. 10.17 Dusky salamander (*Desmognathus fuscus*) crawling through moss in a New York wetland. (Photograph courtesy of Flickr contributor Dave Huth)

Threats to Wetlands and Public Perceptions

Historically in pre-movie times and even into the modern age in films, wetlands have often been portrayed in a negative fashion. Some moviegoers in the middle of the last century were even frightened by the infamous "Creature from the Black Lagoon" in a classic 1954 science fiction movie (Fig. 10.18)! More importantly, wetlands were seen as harboring disease-bearing mosquitoes and thus were often the focus of draining efforts. The major modern threats to wetlands are cattle grazing, row-crop agriculture, extended droughts, human water extraction, intentionally set fires, invasive plant species (such as purple loosestrife *Lythrum salicaria* and the common reed *Phragmites*), and excavation for roads and other human structures. The United Nations "Millennium Ecosystem Assessment" program determined that environmental degradation is more prominent within wetland systems than in any other ecosystem on Earth. Wetlands are often exploited by cities and farming operations where not protected by federal or state regulations, but they can also be constructed as quasi-natural systems for treating municipal and industrial wastewater as well as stormwater runoff and may play a role in water-sensitive urban design. The outlook for the survival of wetlands in North America is better now than it was a half-century ago, but it is important that we continue efforts to protect these vital aquatic systems for nature and humans around the world.

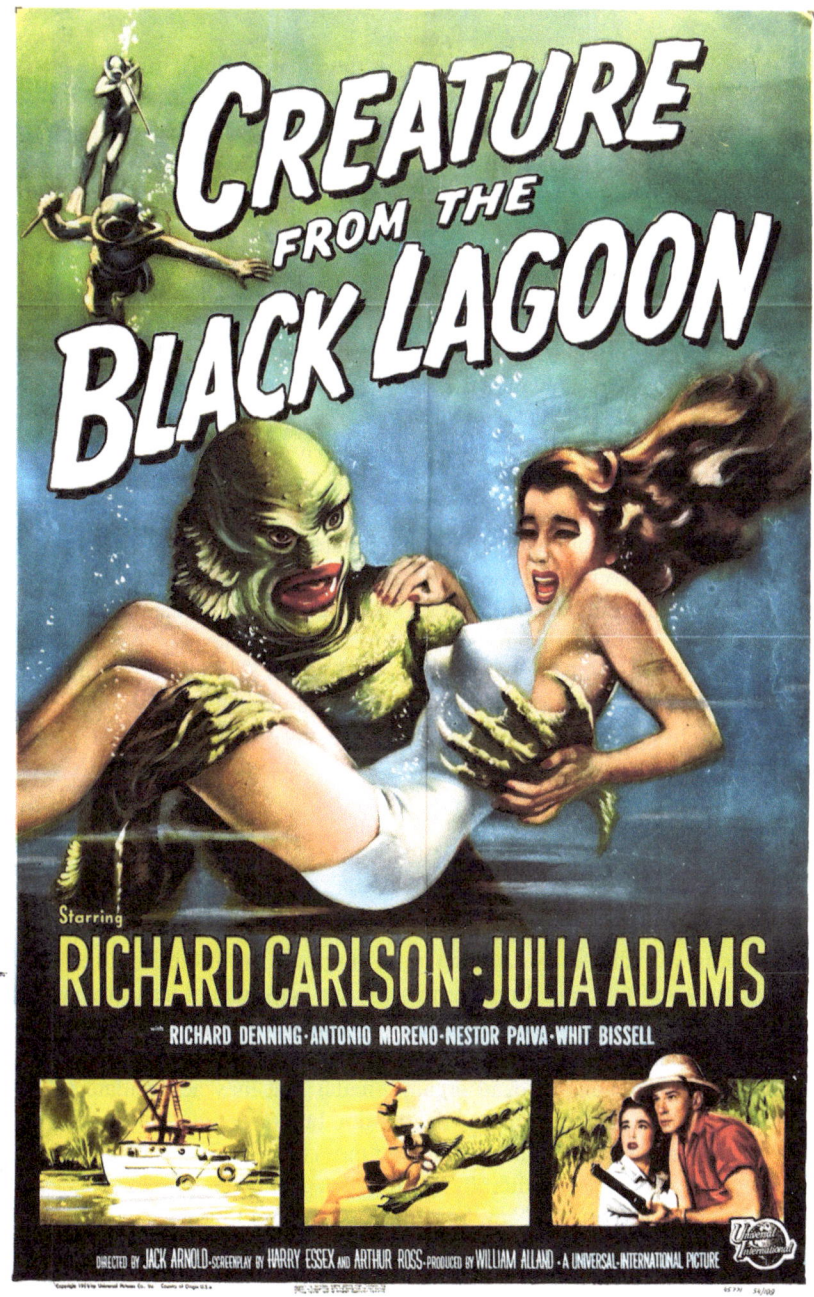

Fig. 10.18 Movie advertisement for the 1954 movie "Creature from the Black Lagoon" which entered the public domain in 1982

What Is So Bizarre About Bizarre Habitats?

Many ecologists prefer to spend their outdoor time in often intriguing but relatively benign and pleasant habitats even though their personal definitions of "pleasant" may differ dramatically from yours given that they may occasionally include working in hot humid forests, or frigid mountain streams, or scorching thorny deserts. A typical attraction of those habitats for ecologists is the opportunity they afford to investigate intriguing research processes or to search for rare plants or animals in these sometimes dangerous habitats. One common characteristic of those seemingly bizarre environments is that the nature of the flora and fauna is often highly distinct from what most of us encounter in our daily lives. Sometimes the species diversity and abundance are high in those habitats (as in tropical forests), but at other times the diversity may be elevated while the abundance is depressed (as in many deserts). Rarely are both the density and diversity of animals low, as in some of the habitats described in this chapter. The habitats discussed below are almost exclusively from the water surface to the bottom of a lake or stream. But for some species like a tropical tree frog, its arboreal aquatic nursery pool is elevated far aboveground to reduce threats to its offspring from many potential predators.

Oh, and why are some of those habitats treacherous? Well, if a human or other animal unexpectedly falls into some of those habitats (like thermal pools or various rivers with large carnivores or massive herbivores), they may never walk again on dry land! An example of the former is the hot springs in Yellowstone National Park, which each year seems to attract some truly foolish and sometimes short-lived tourists and which has accounted for more than 20 deaths of tourists over the years.

The Good, Bad, and Ugly of Bogs

Bogs (Fig. 10.19) are one of the four main types of wetlands (marshes, swamps, bogs, and fens, with the latter two being types of mires). In various English-speaking countries, they are also called mires, mosses, muskegs, quagmires, quaking bogs, and various other names, but here I will group them under the name "bog." These wetlands are characterized by acidic waters, sometimes abundant living aquatic plants, and large amounts of dead organic matter that may accumulate in poorly oxygenated layers from the surface to the bottom. In many cases, this organic matter is derived from mosses, especially sphagnum moss (Fig. 10.20a, b) which can be so thick that you can sometimes literally "walk on water." For example, years ago as a former professor at a private New York university close to the Canadian border, I delivered an ecology field trip lecture while literally standing on a thick layer of moss atop the waters of a bog. Things were going well until a sinking feeling persuaded me that the students enrolled in my class had crowded around me a bit too closely. At that point, a rapid dispersal of the students seemed prudent before more than just the soles of our boots got wet!

Fig. 10.19 A bog with some of the open areas evident. (Photograph courtesy of Christopher F. Frazier)

Fig. 10.20 Sphagnum moss bog with distant (**a**) and close-up (**b**) views. (Photographs courtesy of Christopher F. Frazier)

As implied by the title of this section, not all bogs are bad, despite their occasional unsavory reputation as graveyards (see section below on "Bog Bodies"), because some produce fuel and food for humans and ecosystem services for rare and endangered species found nowhere else. For example and from the human perspective, where we would be without highly nutritious cranberries throughout the year but especially during fall and winter holidays? These fruits of an evergreen dwarf shrub are now raised during part of their life cycle in bogs (Fig. 10.21).

Fig. 10.21 Cranberries being harvested from a bog near Yarmouth, Massachusetts. (Photograph courtesy of Flickr contributor at the Massachusetts Office of Travel and Tourism)

Indigenous people of Canada and the northeastern part of the USA have seasonally consumed cranberries for centuries, often as part of their preparation of pemmican (a mixture of dried berries, meat, and tallow fat), and they later introduced these valuable and long-lasting fruit-containing foods to immigrant Europeans.

Acidic Bogs

The "peat" found in some wetlands is a spongy form of dead organic matter that is a precursor to lignite, soft coal, and eventually hard coal. It is derived from formerly living aquatic vegetation like mosses that can accumulate in the bog's anoxic bottom waters. Peat has traditionally been used over the centuries as a relatively cheap source of energy in many northern latitude countries, and it is now also sold as a soil amendment in some global regions. The latter use, however, is now being discouraged because of the negative ecological effects of draining and excavation of wetlands and the loss of a beneficial habitat that is characterized by carbon sequestration in a time of rapid and damaging climate change.

The unusual environmental conditions characterizing most acidic bogs limit the types of species present. While many birds can exploit these wetlands, along with transitory mammals (e.g., the North American moose; Fig. 10.22) and a few species of frogs, turtles, and lizards, fish are rare in these habitats, probably in part because of the characteristically low oxygen levels resulting from high organic

Fig. 10.22 A moose (*Alces americanus*) wading in a stream. (Photograph courtesy of Flickr contributor Pedro Szekely)

decomposition and oxygen consumption by plants at night. Some highly specialized plants can thrive in bogs with their low nutrients, acidic pH, and permanent or frequent water submersion. These include some small and unusual carnivorous plants, such as sundews (*Drosera*; Fig. 10.23) and pitcher plants (e.g., *Sarracenia*; Fig. 10.24).

Stinky Water

Several natural factors may cause waters to be perceived as "stinky," but perhaps the most common source is linked to methane gas (CH_4). Although aquatic food webs are most often portrayed correctly as being fueled primarily by some combination of nutrients from algae and terrestrial leaf and grass litter, this dichotomy usually ignores a potentially important third pathway to fuel these systems: anaerobic (= without oxygen) metabolism. This chemical process, called methanogenesis, is common at the bottom sediments of ponds, lakes, wetlands, and sometimes side channels of streams where oxygen is absent or extremely low. Aside from the disagreeably smelly aspects of this process, the methane can be converted by methane-oxidizing bacteria (= methanotrophs) through chemical synthesis and then via single-cell protists and other organisms into consumable biomass that helps fuel the benthic invertebrates present in oxygenated portions of the water body. As a

Fig. 10.23 A climbing sundew (*Drosera macrantha*), a carnivorous wetland plant. (Photograph courtesy of Flickr contributor patrickkavanag)

Fig. 10.24 A carnivorous pitcher plant common to many wetlands. (Photograph courtesy of Christopher F. Frazier)

cautionary note, where you smell this methane, that could indicate an ecological situation where decomposition of dead organic matter (= detritus) is slow in general, oxygen is sometimes lethally limited at night, and most fish cannot survive.

Bog Bodies

Perhaps the creepiest thing about bogs (or coolest depending on your perspective) is the presence of "bog bodies" or cadavers that have been mummified over periods of up to 8000 years by the acidic nature of raised peat bogs. These result from the combined effects of water chemistry and low decomposition rates. Bogs are common in northwestern Europe, Canada, and many other places, including in the US state of Florida where nearly 200 preserved bodies have been found, including a 7000-year-old body of a Native American. One place to see excellent displays of bog mummies is in Dublin at the National Museum of Ireland, but you might want to go there during the day because at night they may be too "petrifying"! Unlike dead bodies in terrestrial and non-bog aquatic habitats, some of these human remains have retained their bones, skin, and internal organs because of the chemical nature of the surrounding water, while other remains are limited mostly to the skeleton. Why did these humans end up in a peat bog, if not by accident? There are various speculations, but the most common hypotheses are that these mostly Bronze Age bodies were either "human sacrifices to the gods," slain women perceived as "witches," murder victims, or executed criminals.

These human cadavers—as well as any other animals trapped in the bog—survived mostly to partially because of the acidic characteristic of bogs. The sphagnum moss and peat also release both humic acids and a complex sugar known as sphagnan, which together leach calcium from the bodies and slow decay.

Salt of the Earth

If asked where you should you go to collect animals living in salty waters, your answer would almost invariably be to visit a marine habitat, such as an estuarine salt marsh, or along coastal rocky or sandy intertidal shores, or perhaps to a coral reef. Most likely it would not occur to you to focus on an inland salt lake unless you live in certain regions of the world, such as the US state of Utah where the "Great Salt Lake" is located. While marine habitats other than estuaries (whose salinities fluctuate greatly and daily with the tides) typically have a relatively constant salinity of 35 parts salt dissolved in 1000 parts water (3.5%), freshwater habitats are generally less than one-tenth of 1% saline. Even that level in freshwater is considered "hard water" by most homeowners and would invariably lead to installation of a water softener in your house. By contrast, the hypersaline Great Salt Lake is saltier than the ocean, with its waters commonly ranging from 5% to 27% salt. One of the saltiest marine seas is the semi-confined Red Sea of the Middle East with its 4% salinity. For comparison, the inland "Salton Sea" of California currently has a long-term average of about 4.4% salt (variable with freshwater inputs vs. evaporation), while the inland Dead Sea of Israel and Jordan is 4.5% salt. The Caspian Sea, which is the largest single inland body of water in the world, and which is surrounded by the countries of Azerbaijan, Iran, Kazakhstan, Russia, and Turkmenistan, averages about 1.2% salt.

In addition to this contrast in salt concentrations between marine and the typical freshwater habitats, the chemical nature of the salt is different. That is, the salt in inland water bodies is often not the same substance that you add to your food! For example, the "salt" of the ocean is primarily sodium chloride—which is mostly the same compound that you cook with or add to your food at the dinner table—but the "salt" found in both relatively fresh and intensely salty inland waters can be of substantially different composition and is sometimes toxic. For example, the waters of the Great Salt Lake—a lentic system which historically was derived from an inland sea—consist primarily of the elements sodium and chloride but may also contain substantial amounts of calcium, magnesium, potassium, and sulfate. Other rare inland ponds and lakes may contain dangerous elements, including small amounts of boron, arsenic, and mercury.

For saline inland systems whose waters are not otherwise too toxic or acidic, the diversity of living aquatic organisms that are permanent residents is exceptionally low and may consist mostly of various prokaryotic organisms (e.g., different types of bacteria, including "blue-green algae" or more properly named cyanobacteria), true green algae (eukaryotic, with mostly *Dunaliella*), protozoa, and some multicellular invertebrates. Among the latter are vast numbers of rotifers, brine shrimp (e.g., *Artemia franciscana*), and brine flies (primarily *Ephydra cinerea* in Utah's Great Salt Lake). While the permanent invertebrate inhabitants of these lakes are consistently low in diversity, many resident birds and waterfowl migrating along the Pacific Flyway feed voraciously on the brine flies and fairy shrimp living within these lakes. In fact, some bird species molt their flight feathers in Mono Lake of California, before regrowing them to continue their southward migration. Their survival thus depends substantially on the meals they make of the tiny creatures inhabiting these salty habitats.

Some inland ponds with exceptionally hard water may still support a substantial diversity of normally freshwater animals, like dragonflies (Fig. 10.25). If the water levels drop too low, the residents either flee or die. This can be seen in the protected ponds at the Bitter Lake National Wildlife Refuge in southeastern New Mexico.

Coastal environments over much of the world are often replete with various species of marine barnacles. However, it might surprise you to know that the inland Salton Sea northeast of San Diego, California (Fig. 10.26), includes populations of the marine barnacle *Balanus amphitrite* (Fig. 10.27). Actually, the Salton Sea was originally a freshwater body of water until desiccation from climate change and both water diversion and mining by humans turned it salty over many decades. Many of the few marine creatures found there were introduced intentionally or

Fig. 10.25 Eastern ringtail dragonfly adult (*Erpetogomphus designatus*). (Photograph courtesy of Christopher F. Frazier)

Fig. 10.26 "Salton Sea" in California. (Photograph courtesy of Flickr contributor Kevin Dooley)

accidentally by humans. For example, barnacles seemed to have reached there accidentally in the mid-1940s, possibly via military seaplanes or marine buoys brought from the nearby Pacific Ocean that contained barnacles attached to their exterior.

Some Like It Hot, Others Play It Cool

We tend to think about freshwater organisms existing in waters at temperatures comparable to the surrounding habitats, but some survive in unusually cold or warm temperatures. Cold habitats usually reflect effects of season, latitude (temperate and polar zones), or elevation (high mountains), while warmer temperatures may be

Fig. 10.27 Shells of striped acorn barnacles (*Balanus amphitrite*) found on Cayo Costa Island, Florida. A similar subspecies exists in the inland Salton Sea of California. (Photograph courtesy of Flickr contributor James St. John)

seasonally variable from solar inputs or relatively permanent from latitudinal position or even subterranean waters that continually boil to the surface. Temperatures can also vary with depth in water bodies, especially in lakes (see Chap. 7).

Different kinds of organisms face these external temperatures in often contrasting ways. Internal temperatures of invertebrates and nonmammalian vertebrates like fish, amphibians, and reptiles are poikilothermic ectotherms (a tongue-twisting scientific term for "cold-blooded animals"), and thus their body temperatures fluctuate substantially with external temperatures. In contrast, mammals and birds are endothermic homeotherms ("warm-blooded" animals), and hence their body temperatures remain very constant over time in most species. Aquatic species face temperature extremes in different ways, including migrating to more acceptable climates if possible, burrowing into more constant temperature mud, transitioning into a different life history stage (such resistant, fertilized "eggs"), or entering biotic states of torpor or dormancy through estivation (warm temperatures) or hibernation (cold temperatures).

In Hot Water

If you have ever been told—probably by a parent or boss—that "you are in hot water now," then you know it is not a desirable situation! In fact, few if any animals live in thermal springs (Fig. 10.28) or in outflowing streams or semi-attached thermal pools, at least until the temperature cools to just warm water. Within the hot springs themselves, only "thermophilic" organisms exist, and these consist of bacteria and archaea (= truly ancient types of bacteria). The latter were once considered bacteria but are now separated taxonomically because archaea lack nuclei and have unusual membranes. Archaea-like *Sulfolobus* have been collected from thermal pools at temperatures just below boiling (194 °F/90 °C).

Humans should always be cautious around these hot and warm habitats in general (if not entirely avoiding them) for various reasons including that some contain thermophilic species dangerous to humans, such as the single-cell amoeba *Naegleria* which can cause fatal brain infections. In contrast, warm pools without killing temperatures but present somewhat near the extreme heated areas may attract numerous multicellular, freshwater animals including various invertebrates, a few fish, and often migrating birds. More bizarre are the Japanese macaque (*Macaca fuscata*) or "snow monkeys" who frequently exploit high mountainous thermal springs in winter to survive in these far northern latitudes while gaining a little time for "rest and recuperation"!

High temperatures are not always the direct cause of death of aquatic species. The dangerous aspect for aquatic species can be a lack of oxygen because the concentrations drop precipitously with higher temperatures. Somewhere around 86–90 °F (30–32 °C), river fish will start dying from a lack of sufficient oxygen. A contributing problem in lakes is that algae absorb oxygen at night, so even if the

Fig. 10.28 A hot spring in Yellowstone National Park, USA. (Photograph courtesy of Flickr contributor paweesit)

temperature does not reach otherwise dangerous levels, fish kills may result in warm weather from insufficient oxygen.

Those of us living in the north and south temperate zones often envision the tropics as very hot environments. In fact, while they may feel much hotter because of high humidity, actual temperatures in the tropics outside of cities rarely reach 100 °F/38 °C, in contrast to the often seasonally warmer temperatures in the temperate zone. Consequently, aquatic species living in the temperate zone may be exposed more often to lethal summer temperatures than those in the tropical regions.

Colder Than a Gravestone in Winter

On the opposite thermal extreme are cold habitats in far northern and southern latitudes or in high mountains. Temperatures in water change very slowly because of its high heat capacity which demands a lot of energy in joules (>4000) to change ~2.2 lb/1 kg of water even one degree centigrade (1.4 °F/−17 °C). This characteristic of water is one major reason life can exist on Earth. However, pools and streams in temperate and polar environments freeze seasonally either completely or just near the upper surface where ice floats (see Chap. 7). In higher latitudes, temperatures often range from cool to literally freezing, while in midlatitudes of the temperate zone, temperatures show a greater range from extremely hot (>100 °F/38 °C) to extremely cold (below 0 °F/-18 °C). Freshwater pools and lakes in very high latitudes often freeze entirely from the upper surface to the bottom, making life there virtually impossible for fish and many invertebrates.

Aquatic animals living over many generations in latitudes where waters freeze have different adaptations for survival. Invertebrates in those habitats may enter a life stage (cyst) that can tolerate freezing and later hatch out when the system thaws. This is a strategy followed by chironomid midges in arctic ponds. No amphibians and reptiles can tolerate whole-body tissue freezing, but a few rare species can survive having two-thirds of their body water frozen for short periods. For example, subarctic populations of adult wood frogs (*Rana sylvatica*) can remain frozen for more than half a year because of high concentrations of "antifreeze glycolipids" in their blood. Organic osmolytes (like glucose and urea) in their blood are slower to freeze, thereby providing more habitat flexibility. Even more tolerant are "ice algae" which can survive "within" the surface ice of lakes and rivers as long as they can obtain 0.1% of the surface photic radiation. In general, however, exposure to surface ice slows metabolism, while the complete freezing of a water body typically requires animals either to enter a resistant life stage, survive in a protected surface burrow (in the bank or a constructed aquatic lodge, as in beavers), or migrate to warmer climes where conditions are more favorable.